AGRICULTURE ISSUES AND POLICIES

AGRICULTURAL POLICY, APPROPRIATIONS AND DISCRETIONARY SPENDING

AGRICULTURE ISSUES AND POLICIES

Additional books and e-books in this series can be found on Nova's website under the Series tab.

AGRICULTURE ISSUES AND POLICIES

AGRICULTURAL POLICY, APPROPRIATIONS AND DISCRETIONARY SPENDING

ALBERTUS BRINKHUIS
EDITOR

snova
New York

Library of Congress Cataloging-in-Publication Data

ISBN: 978-1-53615-215-9

Published by Nova Science Publishers, Inc. † New York

CONTENTS

PREFACE

The Agriculture, Rural Development, Food and Drug Administration, and Related Agencies appropriations bill provides funding for a wide array of Federal programs, mostly in the U.S. Department of Agriculture (USDA). These programs include agricultural research, education, and extension activities; natural resources conservation programs; farm income and support programs; marketing and inspection activities; domestic food assistance programs; rural housing, economic and community development, and telecommunication and electrification assistance; and various export and international activities of the USDA. Agriculture appropriations include both mandatory and discretionary spending. Discretionary amounts, though, are the primary focus during the bill's development. This book provides information on the 2018 and 2019 discretionary spending as well as a copy of the 2019 Agriculture appropriations act.

Chapter 1 - This is an edited, reformatted and augmented version of Senate 115th Congress, 2nd Session, Calendar No. 438; REPORT 115–259, dated May 24, 2018.

Chapter 2 - The Agriculture appropriations bill funds the U.S. Department of Agriculture (USDA) except for the Forest Service. It also funds the Food and Drug Administration (FDA) and—in even-numbered fiscal years—the Commodity Futures Trading Commission (CFTC).

Agriculture appropriations include both mandatory and discretionary spending. Discretionary amounts, though, are the primary focus during the bill's development. The largest discretionary spending items are the Special Supplemental Nutrition Program for Women, Infants, and Children (WIC); agricultural research; FDA; rural development; foreign food aid and trade; farm assistance programs; food safety inspection; conservation; and animal and plant health programs. The main mandatory spending items are the Supplemental Nutrition Assistance Program (SNAP), child nutrition, crop insurance, and the farm commodity and conservation programs paid by the Commodity Credit Corporation. For FY2019, both the House and Senate Appropriations Committees reported agriculture bills (H.R. 5961, S. 2976) in May 2018. The Senate amended and passed its version as Division C of a four-bill minibus on August 1, 2018 (H.R. 6147). In the absence of a final appropriation, Congress enacted two continuing resolutions through December 21, 2018 (P.L. 115-245, Division C; and H.J.Res 143). The Trump Administration requested $17 billion for discretionary-funded accounts within the jurisdiction of Agriculture appropriations, a reduction of $6.2 billion from FY2018 (-27%). Both the House-reported and Senate-passed bills generally reject most of the proposed reductions. The discretionary total of the House-reported bill is $23.23 billion, which would be $14 million less than enacted in FY2018 (-0.1%). The discretionary total of the Senate-passed bill is also $23.23 billion. However, the Senate bill's total would be $229 million more than enacted in FY2018 (+1%) on a comparable basis that excludes the CFTC. The Senate-passed bill would provide about $250 million more than the House-passed bill on a comparable basis. The primary changes at the agency level that comprise the differences between the bills and from FY2018 include the following: Both the House and Senate bills would increase FDA appropriations (+$308 million in the House bill; +$159 million in the Senate bill). Both bills increase appropriations for animal and plant health programs (+$16 million to +$19 million). The House bill would provide more base funding for rural water and waste disposal (+$81 million), but none of the extra money that was provided separately in FY2018. The Senate bill would not change the base funding for rural water but continues $400 million of the extra funding from last year. For rural broadband, both the House and Senate bills would continue extra funding from a FY2018 pilot ($550 million in the House bill; $425 million in the Senate bill). The House bill would increase appropriations for agricultural research (+$79 million), and the Senate bill would increase Agricultural Research Service programming (+98 million) but would not provide any money for construction (-$141 million). Both bills provide less for WIC (-$175 million in the House bill, and -$25 million in the Senate bill), though the Senate bill has a larger rescission from prior-year WIC funds than does the House bill. The House bill would reduce base funding for the international Food for Peace program (-$100 million) and does not renew extra funding provided last year (-$116 million), while the Senate bill would keep it constant overall. The appropriations also carry mandatory spending—largely determined in separate authorizing laws—that would total $122 billion. Thus, the overall total of the both bills is about $145 billion. Both bills contain policy provisions affecting disaster programs, rural definitions, industrial hemp, animal products, nutrition programs, dietary guidelines, CFTC, and tobacco products.

Chapter 3 - The Agriculture appropriations bill funds the U.S. Department of Agriculture (USDA) except for the Forest Service. It also funds the Food and Drug Administration (FDA) and—in even-numbered fiscal years—the Commodity Futures Trading Commission (CFTC). Agriculture appropriations include both mandatory and discretionary spending. Discretionary amounts, though, are the primary focus during the bill's development, since mandatory amounts are generally set by authorizing laws such as the farm bill. The largest discretionary spending items are the Special Supplemental Nutrition Program for Women, Infants, and Children (WIC); agricultural research; FDA; rural development; foreign food aid and trade; farm assistance programs; food safety inspection; conservation; and animal and plant health programs. The main mandatory spending items are the Supplemental Nutrition Assistance Program (SNAP), child nutrition, crop insurance, and the farm commodity and conservation programs paid by the Commodity Credit Corporation. Congress passed the FY2018 Consolidated Appropriations Act on March

23, 2018 (P.L. 115-141). The discretionary total of the FY2018 Agriculture appropriation (Division A) is $23.3 billion. This is an increase of $2.1 billion above the amount enacted in FY2017 (+10%). The appropriations also carry nearly $123 billion of mandatory spending, though that amount is largely determined in separate authorizing laws. Thus, the overall total of the enacted FY2018 Agriculture appropriation is about $146 billion. Compared to FY2017, the enacted appropriation increases spending primarily through an extra $1.38 billion in General Provisions, including an additional $500 million for rural water programs, $600 million for rural broadband, $116 million for Food for Peace foreign food aid, and $94 million for opioid enforcement and surveillance. The Consolidated Appropriations Act also increases regular funding for agricultural research by $138 million, and Food for Peace by $134 million. Unlike the past decade, the FY2018 Agriculture appropriation does not include as many changes to mandatory program spending (CHIMPS), such as to the Environmental Quality Incentives Program (EQIP). Both the House and Senate Appropriations Committees reported Agriculture appropriations bills in July 2017 (H.R. 3268, S. 1603). As the beginning of the fiscal year approached, the House passed a consolidated bill in September 2017 that included an agriculture appropriation (Division B of H.R. 3354). The full Senate did not consider an agriculture appropriations bill on the floor. The government continued to operate for the first six months of the fiscal year under continuing resolutions (CRs). The last CR, the Bipartisan Budget Act of 2018 (P.L. 115-123), also enacted supplemental appropriations that included agricultural assistance, amended certain farm bill statutes, and passed new, higher budget caps that facilitated the final appropriation.

Chapter 4 - The Agriculture appropriations bill funds the U.S. Department of Agriculture (USDA) except for the Forest Service. The FY2018 Consolidated Appropriations Act (P.L. 115-141, Division A), and both of the FY2019 agriculture bills reported by the House and Senate Appropriations Committees (H.R. 5961, S. 2976) include funding for conservation programs and activities at USDA. Congress passed the FY2018 Consolidated Appropriations Act on March 23, 2018, which included agriculture appropriations under Division A. For FY2019, the

House and Senate Appropriations Committees reported agriculture bills in May 2018. The Senate amended and passed its version as Division C of a four-bill minibus on August 1, 2018 (H.R. 6147). In the absence of a final appropriation, Congress enacted a continuing resolution through December 7, 2018 (P.L. 115-245, Division C). Agricultural conservation programs include both mandatory and discretionary spending. Most conservation program funding is mandatory and is authorized in omnibus farm bills. Other conservation programs—mostly technical assistance—are discretionary and are funded through annual appropriations. The largest discretionary program is the Conservation Operations (CO) account, which funds conservation planning and implementation assistance on private agricultural lands across the country. The enacted FY2018 appropriation provided $874 million for CO, an increase from the FY2017 enacted amount ($864 million). The FY2019 House-reported and Senate-passed bills would further increase funding for CO above FY2018 levels to $890 million and $879 million, respectively. Other discretionary spending is primarily for watershed programs. The largest—Watershed and Flood Prevention Operations (WFPO)—was funded at $150 million in FY2018. Both the House-reported and Senate-passed bills would fund WFPO at the $150 million level in FY2019. Most mandatory conservation programs are authorized in omnibus farm bills and do not require an annual appropriation. However, Congress has reduced mandatory conservation programs through changes in mandatory program spending (CHIMPS) in the annual agricultural appropriations law every year since FY2003. The enacted FY2018 omnibus marks the first appropriation since FY2002 that does not include CHIMPS to mandatory conservation programs. For FY2019, both the House and Senate appropriation bills do not include reductions to mandatory conservation programs, because most programs' authorizations expired on September 30, 2018, making these programs ineligible for reduction. While this is infrequent, the Agriculture appropriations bill may also serve as a vehicle for amendments to authorized programs that permanently alter or create programs. The FY2018 enacted appropriation included two such amendments—one to WFPO and one to farm bill conservation program reporting requirements.

The WFPO amendment increased the size threshold required for congressional approval. Under the amended language, the Senate and House Agriculture Committees must approve WFPO projects that include an estimated federal contribution of more than $25 million for construction, an increase from the previous $5 million threshold. Additionally, the FY2018 appropriation exempted farm bill conservation programs from select federal reporting requirements, including obtaining a Data Universal Numbering System (DUNS) number and System for Award Management (SAM) registration. Agriculture appropriations bills may also include policy-related provisions that direct how the executive branch should carry out the appropriation. The FY2018 enacted appropriation and both the FY2019 House-reported and Senate-passed bills include policy provisions for conservation programs that range from reports to Congress to suggested natural resource priorities.

Chapter 5 - The Agriculture appropriations bill—formally known as the Agriculture, Rural Development, Food and Drug Administration, and Related Agencies Appropriations Act—funds the Food and Drug Administration (FDA) and the U.S. Department of Agriculture (USDA), excluding the U.S. Forest Service. Congress enacted the FY2018 agriculture appropriation in March 2018 as part of the Consolidated Appropriations Act, 2018 (P.L. 115-141, Division A). Both the House and the Senate Appropriations Committees have reported Agriculture appropriations bills for FY2019 (H.R. 5961, S. 2976). The Senate amended and passed its version as Division C of a four-bill minibus (H.R. 6147). Numerous federal, state, and local agencies share responsibilities for regulating the safety of the U.S. food supply. Federal responsibility for food safety rests primarily with FDA, an agency of the Department of Health and Human Services, and also the Food Safety and Inspection Service (FSIS), an agency of USDA. FDA is responsible for ensuring the safety of the majority of all domestic and imported food products—except for meat and poultry products, which are within USDA's jurisdiction to oversee meat, poultry, and processed egg products. Combined appropriations covering food safety activities at both FDA and USDA totaled nearly $2.1 billion in FY2018. Congressional appropriations at both

FDA and USDA are augmented by existing (currently authorized) user fees. FDA user fees authorized by the FDA Food Safety Modernization Act (FSMA, P.L. 111-353) have generated between $10 million and $18 million annually in recent years. At FSIS, user fees have generated between $180 million and $250 million per year. At FDA, ongoing efforts to improve food safety include implementation of FSMA. FSMA was enacted by the 111th Congress and was the largest expansion of FDA's food safety authorities since the 1930s. Since FSMA became law in 2011, congressional appropriators have increased annual funding for the FDA Foods Program by $204.3 million—an increase of about 24% between FY2011 and FY2018—largely in an effort to support FDA's implementation of FSMA. The enacted FY2018 appropriation for FDA's Foods Program provided $1,041.6 million. Currently, FDA funding for its food safety oversight activities is roughly similar to those of FSIS (*Figure 1*). FDA's total budget for food safety programs and activities extends beyond the agency's Foods Program, encompassing other food and veterinary medicine programs at FDA. Food-safety-related activities at FSIS include continuous inspections at federal meat and poultry plants. The enacted FY2018 appropriation provided $1,056.8 million to carry out this function. Compared to FY2011—the year FSMA was enacted—annual congressional appropriations for FSIS have increased by $48.3 million (+5%). For FY2019, congressional appropriators would increase funding for federal food safety activities, whereas the Administration's budget proposal would reduce food safety funding below FY2018 levels at both FDA and FSIS.

In: Agricultural Policy ...
Editor: Albertus Brinkhuis

ISBN: 978-1-53615-215-9

Chapter 1

AGRICULTURE, RURAL DEVELOPMENT, FOOD AND DRUG ADMINISTRATION, AND RELATED AGENCIES APPROPRIATIONS BILL, 2019*

U.S. Senate

May 24, 2018

Mr. HOEVEN, from the Committee on Appropriations, submitted the following:

REPORT

[To accompany S. 2976]

The Committee on Appropriations reports the bill (S. 2976) making appropriations for Agriculture, Rural Development, Food and Drug Administration, and Related Agencies programs for the fiscal year ending

* This is an edited, reformatted and augmented version of Senate 115th Congress, 2nd Session, Calendar No. 438; REPORT 115–259, dated May 24, 2018.

September 30, 2019, and for other purposes, reports favorably thereon and recommends that the bill do pass.

New obligational authority	
Total of bill as reported to the Senate	$145,446,786,000
Amount of 2018 appropriations	151,145,985,000
Amount of 2019 budget estimate	139,317,668,000
Bill as recommended to Senate compared to—	
2018 appropriations	-5,699,199,000
2019 budget estimate	+6,129,118,000

BREAKDOWN BY TITLE

The amounts of obligational authority for each of the seven titles are shown in the following table. A detailed tabulation, showing comparisons, appears at the end of this chapter. Recommendations for individual appropriation items, projects and activities are carried in this chapter under the appropriate item headings.

[In thousands of dollars]		
	Fiscal year 2018 enacted	Committee recommendation
Title I: Agricultural programs	6,966,837	6,967,783
Title II: Farm Production and Conservation programs	25,933,930	26,825,522
Title III: Rural economic and community development programs	3,000,881	3,000,883
Title IV: Domestic food programs	104,919,418	103,040,913
Title V: Foreign assistance and related programs	2,020,957	2,152,323
Title VI: Related agencies and Food and Drug Administration	2,811,866	2,970,866
Title VII: General provisions	577,096	488,496
Supplemental Appropriations for Disaster	3,645,000	
Total, new budget (obligational) authority	151,145,985	145,446,786

OVERVIEW AND SUMMARY OF THE BILL

The Agriculture, Rural Development, Food and Drug Administration, and Related Agencies appropriations bill provides funding for a wide array of Federal programs, mostly in the U.S. Department of Agriculture [USDA]. These programs include agricultural research, education, and extension activities; natural resources conservation programs; farm income and support programs; marketing and inspection activities; domestic food assistance programs; rural housing, economic and community development, and telecommunication and electrification assistance; and various export and international activities of the USDA.

The bill also provides funding for the Food and Drug Administration [FDA] and allows the use of collected fees for administrative expenses of the Farm Credit Administration [FCA].

The discretionary programs and activities of USDA and FDA that are supported by this bill include high priority responsibilities entrusted to the Federal Government and its partners to protect human health and safety, contribute to economic recovery, and achieve policy objectives strongly supported by the American people. The ability to provide for these measures is made difficult by growing pressure on available levels of discretionary spending as a consequence of the overall public debate on Federal spending, revenues, and size of the Federal debt.

Too often, the USDA programs funded by this bill are confused with farm subsidies and other mandatory spending more properly associated with multi-year farm bills. In contrast, this bill provides annual funding for programs familiar to all Americans such as protecting food safety through the Food Safety and Inspection Service and the Food and Drug Administration, which also plays a vital role in maintaining the safety of the Nation's blood supply and availability of safe and effective medical products and other components of our health system. This bill also provides funding to fight against the introduction and spread of noxious or infectious and often invasive pests and disease that threaten our plant and animal health environments, as well as funding for many other missions of dire importance to the American people.

In the context of overall pressures on spending and the competing priorities that the Committee faces, this bill as reported provides the proper amount of emphasis on agricultural, rural development, and other programs and activities funded by the bill. It is consistent with the subcommittee's allocation for fiscal year 2019. All accounts in the bill have been closely examined to ensure that an appropriate level of funding is provided to carry out the programs of USDA, FDA, and FCA. Details on each of the accounts, the funding level, and the Committee's justifications for the funding levels are included in the report.

Reports to Congress

The Committee has, throughout this chapter, requested agencies to provide studies and reports on various issues. The Committee utilizes these reports to evaluate program performance and make decisions on future appropriations. The Committee directs that all studies and reports be provided to the Committee as electronic documents in an agreed upon format within 120 days after the date of enactment, unless an alternative submission schedule is specifically stated in the report request.

Title I: Agricultural Programs Processing, Research, and Marketing

Office of the Secretary

(Including Transfers of Funds)	
Appropriations, 2018	$46,532,000
Budget estimate, 2019	42,498,000
Committee recommendation	46,532,000

The Secretary of Agriculture, assisted by the Deputy Secretary, Under Secretaries and Assistant Secretaries, Chief Information Officer, Chief Financial Officer, and members of their immediate staffs, directs and coordinates the work of the Department. This includes developing policy, maintaining relationships with agricultural organizations and others in the development of farm programs, and maintaining liaison with the Executive

Office of the President and Members of Congress on all matters pertaining to agricultural policy. The general authority of the Secretary to supervise and control the work of the Department is contained in the Organic Act (7 U.S.C. 2201–2202). The delegation of regulatory functions to Department employees and authorization of appropriations to carry out these functions is contained in 7 U.S.C. 450c–450g.

Committee Recommendations

The Committee recommends an appropriation of $46,532,000 for the Office of the Secretary. The Committee recommendation includes the following accounts under the Office of the Secretary: Office of the Secretary; Office of Tribal Relations; Office of Homeland Security and Emergency Coordination; Office of Advocacy and Outreach; Office of the Assistant Secretary for Administration; Departmental Administration; Office of Assistant Secretary for Congressional Relations; and Office of Communications. The following table reflects the amount provided by the Committee for each office and activity:

[In thousands of dollars]			
	Fiscal year 2018 enacted	Fiscal year 2019 budget request	Committee recommendation
Office of the Secretary	5,051	4,850	5,051
Office of the Assistant to the Secretary for Rural Development	800	800	800
Office of Homeland Security	1,496	1,448	1,496
Office of Partnerships and Public Engagement	4,711	1,672	4,711
Office of Assistant Secretary for Administration	804	875	804
Departmental Administration	22,301	22,501	22,301
Office of Assistant Secretary for Congressional Relations	3,869	3,091	3,869
Office of Communications	7,500	7,261	7,500
Total	46,532	42,498	46,532

Century Farms

The Committee encourages the Secretary to create a program to recognize farms that have been in operation for 100 years and encourages the establishment of a National Century Farms Designation.

Commodity Credit Corporation [CCC] Obligations and Commitments

The Secretary is directed to notify the Committees on Appropriations of the House and Senate in writing 15 days prior to the obligation or commitment of any emergency funds from the CCC.

Livestock Crossing

The Committee is concerned with the ongoing problem of the crossing of livestock from Mexico into the U.S. without proper inspection, which creates risk of disease and loss of forage for U.S. ranchers in the Southwest border region. The Committee directs the agency, in consultation with other Federal and State agencies, to develop a plan of action to better prevent and reduce unauthorized international crossing of livestock on the South-west border.

Multi-Agency Transparency

The Committee expresses support for increasing transparency within all agencies of the Department of Agriculture. The agencies are encouraged to disclose costs associated with analyses required by the National Environmental Policy Act.

Outreach to Socially Disadvantaged and Veteran Farmers and Ranchers

The Committee supports the efforts of the Office of Advocacy and Outreach to increase the accessibility of USDA programs to underserved constituents, and notes that $10,000,000 in mandatory funds is available to assist socially disadvantaged and veteran farmers and ranchers in owning and operating farms and ranches to meet the growing need for financial, production, management, and other assistance to those communities and address workforce shortages. Additionally, the Committee recommendation includes $3,000,000 in discretionary funding for these activities.

Resource Conservation and Development Councils

Since 1964, the Resource Conservation and Development [RC&D] Councils have worked at the grassroots level with local leaders to plan, develop, and carry out programs for land and water conservation and management. The Committee encourages the Secretary to consider the maximum practical use of RC&D Councils, where such RC&D Councils meet agency performance requirements, in the delivery of USDA programs and services.

State Rural Development Councils

The Committee recognizes the successful work of State Rural Development Councils [SRDCs) and their role in advancing rural America and promoting strength and prosperity across the country, and urges the Secretary to provide resources to help improve and expand the impact of SRDCs.

Urban Agriculture

The Committee is aware of a steady increase in urban agriculture initiatives taking place in metropolitan areas across the country. The Committee strongly supports such initiatives and recognizes that successful, robust urban farms can positively impact urban communities and residents in a variety of ways by providing education, entrepreneurial opportunities, and job training; addressing shortages of fresh fruits and vegetables; increasing health and wellness of pregnant women and young children; and reducing obesity rates, recidivism, and urban blight. The Committee commends the Department's efforts to foster such initiatives and encourages the Secretary to increase support and outreach for urban agriculture, including grants, loans, and technical assistance for these innovative urban horticulture projects.

Wheat Grading

The Committee is concerned about unfair wheat grading practices that negatively affects American wheat growers that export to Canada. Current Canadian wheat grading law automatically downgrades America wheat to the lowest quality designation regardless of the type or quality of the wheat. In the United States, however, our grading system provides a fair examination for wheat imported from Canada. This discrepancy needs to be addressed to ensure our wheat growers are being treated fairly. Therefore, the Committee urges the Secretary of Agriculture to work with the Department of Commerce and the United States Trade Representative to prioritize initiating conversations with the Canadian Government to address trade inequities resulting from Canada's current wheat grading practices.

Zoonotic Disease Collaboration

The Committee believes that complex problems affecting the health of humans, animals, and the environment are best solved through important communication, cooperation, and collaboration across disciplines, sectors, between agencies, and between other appropriate domestic and international actors. The Committee directs USDA to provide a report within 60 days of enactment of this Act detailing existing collaborative efforts between FDA, USDA, and other agencies to prevent and respond to zoonotic disease outbreaks in animals and humans; a proposed framework to improve these efforts; and specific activities requested to achieve the proposed framework.

Executive Operations

Executive operations were established as a result of the reorganization of the Department to provide a support team for USDA policy officials and selected department-wide services. Activities under the executive operations include the Office of the Chief Economist, the National Appeals Division, and the Office of Budget and Program Analysis.

Office of the Chief Economist

Appropriations, 2018	$19,786,000
Budget estimate, 2019	19,487,000
Committee recommendation	19,786,000

The Office of the Chief Economist advises the Secretary of Agriculture on the economic implications of Department policies and programs. The Office serves as the single focal point for the Nation's economic intelligence and analysis, risk assessment, and cost-benefit analysis related to domestic and international food and agriculture issues, provides policy direction for renewable energy development, conducts analyses of climate change impacts on agriculture and forestry, and is responsible for coordination and review of all commodity and aggregate agricultural and food-related data used to develop outlook and situation material within the Department.

Committee Recommendations

The Committee recommends an appropriation of $19,786,000 for the Office of the Chief Economist.

Policy Research

The Committee recommendation includes $4,000,000 for policy research under 7 U.S.C. 3155 for entities with existing institutional capacity to conduct complex economic and policy analysis and a lengthy and well-documented record of conducting policy analysis for the benefit of the Department of Agriculture, the Congressional Budget Office, or the Congress. To maximize resources, the Committee expects the Department to focus efforts on entities that have developed models, databases, and staff necessary to conduct in-depth analysis of impacts of agriculture or rural development policy proposals on rural communities, farmers, agribusiness, taxpayers, and consumers.

Office of Hearings and Appeals

Appropriations, 2018	$15,222,000
Budget estimate, 2019	14,183,000
Committee recommendation	15,222,000

The Office of Hearings and Appeals conducts administrative hearings and reviews of adverse program decisions made by the Rural Development mission area, the Farm Service Agency, the Risk Management Agency, and the Natural Resources Conservation Service.

Committee Recommendations

The Committee recommends an appropriation of $15,222,000 for the Office of Hearings and Appeals.

Office of Budget and Program Analysis

Appropriations, 2018	$9,525,000
Budget estimate, 2019	8,631,000
Committee recommendation	9,525,000

The Office of Budget and Program Analysis provides direction and administration of the Department's budgetary functions including development, presentation, and execution of the budget; reviews program and legislative proposals for program, budget, and related implications; analyzes program and resource issues and alternatives, and prepares summaries of pertinent data to aid the Secretary and departmental policy officials and agency program managers in the decisionmaking process; and provides department-wide coordination for and participation in the presentation of budget-related matters to the committees of the Congress, the media, and interested public. The Office also provides department-wide coordination of the preparation and processing of regulations and legislative programs and reports.

Committee Recommendations

The Committee recommends an appropriation of $9,525,000 for the Office of Budget and Program Analysis.

Office of the Chief Information Officer

Appropriations, 2018	$58,950,000
Budget estimate, 2019	62,524,000
Committee recommendation	63,950,000

The Office of the Chief Information Officer was established in August 1996 (40 U.S.C. 1401 et seq.), pursuant to the Clinger-Cohen Act of 1996, which required the establishment of a Chief Information Officer for major Federal agencies. This Office provides policy guidance, leadership, coordination, and direction to the Department's information management and information technology investment activities in support of USDA program delivery, and is the lead office in USDA e-gov efforts. The Office provides long-range planning guidance, implements measures to ensure that technology investments are economical and effective, coordinates interagency information resources management projects, and implements standards to promote information exchange and technical interoperability. In addition, the Office of the Chief Information Officer is responsible for certain activities financed under the Department's Working Capital Fund (7 U.S.C. 2235). The Office also provides telecommunication and automated data processing [ADP] services to USDA agencies through the National Information Technology Center with locations in Fort Collins, Colorado, Kansas City, Missouri and Washington, DC. Direct ADP operational services are also provided to the Office of the Secretary, the Office of the General Counsel, the Office of Communications, the Office of the Chief Financial Officer, and Departmental Management.

Committee Recommendations

The Committee recommends an appropriation of $63,950,000 for the Office of the Chief Information Officer. This includes an increase of $5,000,000 for enhanced cybersecurity activities.

Mobile Derived Credentials

The Committee provides the requested increase for IT Modernization including cyber security initiatives to ensure the Department maintains and increases its cyber security posture. Of the funding provided $1,000,000 is for the procurement and implementation of Mobile Derived Credentials to allow all employees, including field offices, secure and rapid access to all Departmental systems from a mobile and remote environment. The Department shall also use these funds to develop similar mobile and remote access credentials for consumers to easily and securely access Department consumer applications, therefore reducing the need for consumers to travel to field offices for transactions.

Software Licenses

The Committee encourages the Department's Chief Information Officer to perform periodic automated inventories of software licenses in use across the Department. The Department should compare those usage numbers to its purchased licenses and seek to increase efficiency wherever it identifies discrepancies. The Department is to consider using this information to obtain department-wide acquisitions as opposed to component-specific purchases of licenses.

Office of the Chief Financial Officer

Appropriations, 2018	$6,028,000
Budget estimate, 2019	5,536,000
Committee recommendation	6,028,000

The Office of the Chief Financial Officer is responsible for the dual roles of Chief Financial Management Policy Officer and Chief Financial Management Advisor to the Secretary and mission area heads. The Office provides leadership for all financial management, accounting, travel, Federal assistance, and strategic planning performance measurement activities within the Department. The Office is also responsible for the management and operation of the National Finance Center and the Departmental Working Capital Fund.

Committee Recommendations

The Committee recommends an appropriation of $6,028,000 for the Office of the Chief Financial Officer.

Office of the Assistant Secretary for Civil Rights

Appropriations, 2018	$901,000
Budget estimate, 2019	800,000
Committee recommendation	901,000

The Office of the Assistant Secretary for Civil Rights provides oversight of civil rights and related functions. This includes coordination of the administration of civil rights laws and regulations for employees of the Department of Agriculture and participants in programs of the Department, and ensuring compliance with civil rights laws.

Committee Recommendations

The Committee recommends an appropriation of $901,000 for the Office of the Assistant Secretary for Civil Rights.

Office of Civil Rights

Appropriations, 2018	$24,206,000
Budget estimate, 2019	22,345,000
Committee recommendation	24,206,000

The Office of Civil Rights provides overall leadership responsibility for all department-wide civil rights activities. These activities include employment opportunity as well as program nondiscrimination policy development, analysis, coordination, and compliance. The Office is responsible for providing leadership in facilitating the fair and equitable treatment of USDA employees, and for monitoring program activities to ensure that all USDA programs are delivered in a nondiscriminatory manner. The Office's outreach functions provide leadership, coordination, facilitation, and expertise to internal and external partners to ensure equal and timely access to USDA programs for all constituents, with emphasis on the underserved, through information sharing, technical assistance, and training.

Committee Recommendation

The Committee recommends an appropriation of $24,206,000 for the Office of Civil Rights.

Agriculture Buildings and Facilities

(Including Transfers of Funds)	
Appropriations, 2018	$64,414,000
Budget estimate, 2019	58,330,000
Committee recommendation	58,330,000

Department headquarters presently operates in a two-building, Government-owned complex in downtown Washington, DC, the George Washington Carver Center in Beltsville, Maryland, and in leased buildings in the metropolitan Washington, DC, area. Under an arrangement with the General Services Administration, USDA operates, maintains, and repairs these facilities, in lieu of rental payments. For the last several years the Department has implemented a strategic space plan to locate staff more efficiently, renovate its buildings, and eliminate safety hazards, particularly in the Agriculture South Building.

Committee Recommendations

The Committee recommends an appropriation of $58,330,000 for Agriculture Buildings and Facilities.

Hazardous Materials Management

(Including Transfers of Funds)	
Appropriations, 2018	$3,503,000
Budget estimate, 2019	3,463,000
Committee recommendation	3,503,000

Under the Comprehensive Environmental Response, Compensation, and Liability Act and the Resource Conservation and Recovery Act, the Department has the responsibility to meet the same standards regarding the storage and disposition of hazardous materials as private businesses. The Department is required to contain, cleanup, monitor, and inspect for hazardous materials in areas under the Department's jurisdiction.

Committee Recommendations

The Committee recommends an appropriation of $3,503,000 for Hazardous Materials Management.

Office of Inspector General

Appropriations, 2018	$98,208,000
Budget estimate, 2019	87,436,000
Committee recommendation	98,208,000

The Office of Inspector General [OIG] was established October 12, 1978, by the Inspector General Act of 1978 (Public Law 95– 452). This act expanded and provided specific authorities for the activities of OIG which had previously been carried out under the general authorities of the Secretary of Agriculture.

The Office is administered by an inspector general who reports directly to the Secretary of Agriculture. Functions and responsibilities of this Office include direction and control of audit and investigative activities within the Department, formulation of audit and investigative policies and procedures regarding Department programs and operations, and analysis and coordination of program-related audit and investigation activities performed by other Department agencies.

The activities of this Office are designed to assure compliance with existing laws, policies, regulations, and programs of the Department's agencies, and to provide appropriate officials with the means for prompt corrective action where deviations have occurred. The scope of audit and investigative activities is large and includes administrative, program, and criminal matters. These activities are coordinated, when appropriate, with various audit and investigative agencies of the executive and legislative branches of the Government.

Committee Recommendations

The Committee recommends an appropriation of $98,208,000 for the Office of Inspector General. The recommendation also includes funding for OIG to address violations of section 26 of the Animal Welfare Act (7 U.S.C. 2156) and to coordinate with State and local law enforcement personnel in this effort.

Office of the General Counsel

Appropriations, 2018	$44,546,000
Budget estimate, 2019	41,717,000
Committee recommendation	45,146,000

The Office of the General Counsel provides all legal advice, counsel, and services to the Secretary and to all agencies, offices, and corporations of the Department. The Office represents the Department in administrative proceedings; nonlitigation debt collection proceedings; State water rights adjudications; proceedings before the Environmental Protection Agency, Interstate Commerce Commission, Federal Maritime Administration, and International Trade Commission; and, in conjunction with the Department of Justice, in judicial proceedings and litigation.

Committee Recommendations

The Committee recommends an appropriation of $45,146,000 for the Office of the General Counsel. This includes an increase of $600,000 to support international trade activities, as requested in the budget.

Office of Ethics

Appropriations, 2018	$4,136,000
Budget estimate, 2019	2,897,000
Committee recommendation	4,136,000

The Office of Ethics is the centralized and consolidated office implementing USDA's ethics program throughout the Department. The Office provides ethics services to all employees at the Department concerning advice, training, and guidance about compliance with conflict of interest and impartiality rules. This includes complying with the requirements of the Stop Trading on Congressional Knowledge Act, Public Law 112–105 (the STOCK Act), and the Office of Government Ethics regulatory requirements (5 CFR parts 2634 through 2641).

Committee Recommendations

The Committee recommends an appropriation of $4,136,000 for the Office of Ethics.

Office of the Under Secretary for Research, Education, and Economics

Appropriations, 2018	$800,000
Budget estimate, 2019	800,000
Committee recommendation	800,000

The Office of the Under Secretary for Research, Education, and Economics provides direction and coordination in carrying out the laws enacted by the Congress for food and agricultural research, education, extension, and economic and statistical information. The Office has oversight and management responsibilities for the Agricultural Research Service; National Institute of Food and Agriculture; Economic Research Service; and National Agricultural Statistics Service.

Committee Recommendations

The Committee recommends an appropriation of $800,000 for the Office of the Under Secretary for Research, Education, and Economics.

Industrial Hemp

The Committee is aware of statements made by the Department acknowledging the eligibility of researchers participating in industrial hemp pilot programs as defined by Section 7606 of the Agricultural Act of 2014 to compete for Federal funds awarded by the Department. The Committee directs the Department to work with and inform stakeholders of this eligibility and to support industrial hemp research as authorized by Section 7606 of the Agricultural Act of 2014.

Economic Research Service

Appropriations, 2018	$86,757,000
Budget estimate, 2019	45,000,000
Committee recommendation	86,757,000

The Economic Research Service [ERS] provides economic and other social science research and analysis for public and private decisions on agriculture, food, the environment, and rural America.

The information that ERS produces is for use by the general public and to help the executive and legislative branches develop, administer, and evaluate agricultural and rural policies and programs.

Committee Recommendations

The Committee recommends an appropriation of $86,757,000 for the Economic Research Service.

Breastfeeding Study

The Committee recognizes the important role of the Special Supplemental Nutrition Program for Women, Infants, and Children [WIC] in encouraging breastfeeding. Breastfeeding can be an important preventive measure in infant and maternal health, and WIC offers multiple services and supports to mothers to help achieve optimal breastfeeding. As Congress looks for ways to reduce Federal healthcare spending, it is important to understand the preventive impact of breastfeeding and WIC's initiatives within broader healthcare spending. The Committee requests within 12 months an updated study from the ERS on the economic benefits of breastfeeding, including its potential cost-savings for Medicaid and the WIC program.

Feed Costs

The Committee maintains funding provided in fiscal year 2018 for ERS to expand its current feed cost components surveys nationally.

Low Density Polyethylene

The use of Low Density Polyethylene [LDPE] as an agricultural aid on farms is common practice but as a single use material represents a negative impact on the environment. The Committee encourages ERS to conduct research into the viability of creating collection networks and potential markets for agricultural LDPE. The Committee directs NASS to include data from Alaska in compiling the report.

Organic Data Analysis

The organic industry has grown at a tremendous rate over the past several years, and accurate data for the production, pricing and marketing of organic products is essential. Therefore, the Committee encourages ERS to continue and expand the efforts relating to organic data analysis.

National Agricultural Statistics Service

Appropriations, 2018	$191,717,000
Budget estimate, 2019	165,000,000
Committee recommendation	174,767,000

The National Agricultural Statistics Service [NASS] administers the Department's program of collecting and publishing current national, State, and county agricultural statistics. These statistics provide accurate and timely projections of current agricultural production and measures of the economic and environmental welfare of the agricultural sector which are essential for making effective policy, production, and marketing decisions. NASS also furnishes statistical services to other USDA and Federal agencies in support of their missions, and provides consulting, technical assistance, and training to developing countries.

NASS is also responsible for administration of the Census of Agriculture, which is taken every 5 years and provides comprehensive data on the agricultural economy including: data on the number of farms, land use, production expenses, farm product values, value of land and buildings, farm size and characteristics of farm operators, market value of agricultural production sold, acreage of major crops, inventory of livestock and poultry, and farm irrigation practices.

Committee Recommendations

The Committee recommends an appropriation of $174,767,000 for the National Agricultural Statistics Service. This includes an increase of $600,000 for the Geospatial Improvement Initiative as requested in the budget.

Alfalfa Prices

The Committee is concerned that the National Agricultural Statistics Service monthly Agricultural Prices Report only lists the average price received by agricultural producers for alfalfa sold, with no further breakdown of alfalfa hay that meets the higher quality standards required for dairy feed (graded as premium or better, or the equivalent). The Committee directs the National Agricultural Statistics Service to calculate and report in the monthly Agricultural Prices Report on the average price of premium or better alfalfa sold in the United States.

Barley Estimates

The Committee recommends that NASS reinstate acreage and production estimates for barley in States that were discontinued in 2016.

Chemical Use Data Series

The Committee believes that the Chemical Use Data Series provides timely, valuable information on fertilizer and chemical use data on major field crops and selected specialty crops. The Committee encourages the National Agricultural Statistics Service to continue funding the collection and analysis of chemical use data as well as practices such as integrated pest management. The Committee supports the National Agricultural Statistics Service's effort to resume collecting Fruit Chemical Use data and Vegetable Chemical Use data in alternating years and also directs the continuation of this practice to ensure equal access to Federal statistics.

Floriculture Crops Report

The Committee recognizes the importance of the Floriculture Crops Report, and recommends an increase of $500,000 for NASS to complete the report. The Committee directs NASS to include data from Alaska in compiling the report.

Organic Data Initiative

The Committee encourages NASS and AMS to coordinate activities related to expanding organic price reporting and organic data collection, and provides NASS an additional $250,000 for these activities.

Agricultural Research Service

Salaries and Expenses

Appropriations, 2018	$1,202,766,000
Budget estimate, 2019	1,018,991,000
Committee recommendation	1,300,966,000

The Agricultural Research Service [ARS] is responsible for conducting basic, applied, and developmental research through its major program areas of New Products/Product Quality/Value Added; Livestock/Crop Production; Food Safety; Livestock/Crop Protection; Human Nutrition; and Environmental Stewardship. The research applies to a wide range of goals; commodities; natural resources; fields of science; and geographic, climatic, and environmental conditions.

ARS is also responsible for the Abraham Lincoln National Agricultural Library which provides agricultural information and library services through traditional library functions and modern electronic dissemination to agencies of the USDA, public and private organizations, and individuals.

As the USDA's in-house agricultural research unit, ARS has major responsibilities for conducting and leading the national agricultural research effort. It provides initiative and leadership in five areas: research on broad regional and national problems, research to support Federal action and regulatory agencies, expertise to meet national emergencies, research support for international programs, and scientific resources to the executive branch and Congress.

The mission of ARS research is to develop and transfer solutions to agricultural problems of high national priority and provide information access and dissemination to ensure high-quality, safe food and other agricultural products; assess the nutritional needs of Americans; sustain a competitive agricultural economy; enhance the natural resource base and the environment; and provide economic opportunities for rural citizens, communities, and society as a whole.

Committee Recommendations

The Committee recommends an appropriation of $1,300,966,000 for salaries and expenses of the Agricultural Research Service. The Committee does not concur with the President's budget request regarding the termination of research programs and laboratory closures. The Committee expects extramural research to be funded at no less than the fiscal year 2018 levels.

Aerial Application Research

The Committee recognizes the importance of the ARS Aerial Application Technology Program. The program conducts innovative research making aerial applications more efficient, effective and precise. Research for aerial application serves the public good as a vital tool for the future, as agriculture strives to meet the food, fiber, and bio-energy demands of a growing population.

Agricultural Genomics

The Committee provides no less than the fiscal year 2018 level for agricultural genomic research to expand the knowledge of public and private sector entities and persons concerning genomes for species of importance to the food and agriculture sectors in order to maximize the return on the investment in genomics of agriculturally important species.

Agroforestry

Agroforestry can provide on-farm financial and environmental benefits while also addressing the regional and national-scale issues of clean water, wildlife habitat, and hypoxia. Agroforesters manage trees with crops, livestock, and pasture to combine the best of both agriculture and forestry. Recognizing the importance of agroforestry to farm practices and the environment, the Committee recommendation includes no less than the fiscal year 2018 level to develop integrated strategies to manage multifunctional agricultural landscapes that combine trees with agricultural and horticultural crops, forages and grazing livestock for optimal economic, environmental, and natural resources benefits.

Alfalfa Research

The Committee notes that research into alfalfa seed and alfalfa forage systems holds the potential to increase yields, increase milk production, and improve genetics, and the Committee recommendation includes an increase of $1,000,000 to support research focused on alfalfa improvement. Research should focus on using tools to accelerate and enhance existing breeding programs focused on improving yield and quality parameters; developing innovative harvesting and utilization systems; developing new markets for co-products; and quantifying environmental benefits from alfalfa-based systems.

Antimicrobial Research and Development

The Committee recognizes that successful mitigation of antimicrobial resistance in food and agriculture takes coordinated research efforts targeting animals and interaction with the environment. The Committee strongly supports enhanced research efforts to advance the development of alternatives to antibiotics used in animal production. The Committee encourages ARS to examine the role of nutritional alternatives/feed additives containing bioactives and prebiotics that may lead to reduced antibiotic use and boost immune responses in live-stock. The Committee requests that ARS provide an update on this effort in its fiscal year 2020 budget request.

Aquaculture Seedstock

The Committee is concerned that vital seedstock to support the development of aquaculture in Federal waters of the Gulf of Mexico will be sourced from foreign aquaculture producers. Domestic on-land recirculating aquaculture systems are highly capable of producing seedstock to support significant domestic on-land and offshore aquaculture industry growth including through broodstock acquisition and care, spawning, larval culture techniques. Therefore, the Committee encourages USDA to continue working collaboratively with U.S. aquaculture producers and research institutions that specialize in the development of aquaculture technologies and provides an additional $2,000,000 for the development of aquaculture technology that will ensure a steady supply of warm water marine fish seedstock for economic growth of the U.S. aquaculture industry.

Atlantic Salmon Breeding Program

The Committee directs ARS to continue its Atlantic salmon breeding and domestication work. The Committee notes that domestic salmon farms are required to only use strains of salmon that are of North American origin and that these strains need substantial breeding improvement in order to be competitive with strains currently used by foreign producers. The Committee notes that the current ARS Atlantic salmon breeding program lacks a geneticist and supports efforts by the Department to address this need.

Big Data

The Committee is aware of ARS' effort to develop high performance computing infrastructures for modern agricultural research. The Committee provides $9,000,000 to expand ARS' high-speed network, high-performance computing capabilities, big data storage, and modern informatics expertise to meet both current and future needs.

Center for Pollinator Health

The Committee is aware that bees play a crucial role in U.S. agriculture as pollinators, and that continued colony loss poses a serious threat to future food production. While the Committee commends the Department for the steps it has taken to better understand this problem and how to best address it, the Committee is concerned that the maximum benefits of multiagency efforts have yet to be achieved. The Committee provides $3,000,000 for ARS to establish a Center for Pollinator Health in order to provide a central Federal voice on pollinator health. The Committee encourages ARS to collaborate with Federal and land-grant university partners to examine the impact of pesticides, varroa mites, and other potential contributors to bee colony declines.

Ceratocystis Disease

The Committee directs ARS to continue its study of Ceratocystis in the United States, and implement actions and recommendations for response and management pursuant to Senate Report 115–131.

Chronic Wasting Disease

The Committee recognizes the importance of a live test for cervids potentially affected with chronic wasting disease and provides an additional $1,000,000 for research dedicated to the development of such test.

Citrus Greening Disease Research

The Committee commends ARS on its research efforts on citrus greening disease and encourages the agency to continue working to develop methods to reduce transmission, enhance immunity in citrus trees, and work with industry, universities, growers, and other partners to develop effective control mechanisms. The Committee also encourages ARS to coordinate its efforts with the HuanglongbiAng Multi-Agency Coordination [HLB–MAC] group.

Coffee Germplasm

The Committee provides $1,900,000 to complete its report and implement recommendations on including coffee in the National Clonal Germplasm Repository.

Cotton Ginning

The Committee recognizes the importance of pollution abatement, improving fiber quality, ginning efficiency, cotton seed and other byproducts, and provides an additional $500,000 to expand research in cotton ginning and innovation by existing laboratories.

Cover Crops Research and Outreach

The Committee recognizes the importance of developing profitable and practicable cover crop options for use in dairy, grain, and vegetable production systems, including for use in no-till organic systems and as forages. Therefore, the Committee recommendation includes no less than the fiscal year 2018 level for ARS to support research with the purposes of improving measures of soil health and resiliency, varietal development, optimizing dairy forage species combinations, timing and strategies for cover crop seeding and termination, forage integration into organic dairy systems, and mitigation of environmental and extreme rainfall impacts on water quality and soil security for diverse cover crop systems.

Cranberry Research

The Committee recognizes the need for advancements in water conservation, pest control, disease reduction, and fruit quality improvements in cranberry production. The Committee recommendation includes an increase of $1,000,000 for the improvement of cranberry yields, pest management, disease management, and water resource management by developing fields devoted to cranberry research and collection and storage of samples for analysis in appropriate existing laboratory facilities.

Emerging Cereal Rust Diseases

The Committee is aware that emerging cereal rust diseases are a threat to domestic and world food supplies. Therefore, the Department should continue to dedicate funding to speed efforts to combat cereal rust disease, including development of Ug99-resistant wheat varieties.

Feed Enhancement

The Committee recognizes the potential benefits of using Bromoform [CHBR3], currently produced by *Asparagopsis taxiformis* (red seaweed), as a cattle feed enhancement to reduce pollution. The Committee includes an increase of $1,000,000 for the Livestock Nutrient Management Research Unit [LNMRU] to examine the applicability and potential benefits of Bromoform, whether produced by *Asparagopsis taxiformis* or an alternative method, as a cattle feed enhancement.

Floriculture and Nursery Research

The Committee recognizes the economic importance of the floriculture and nursery sector of agriculture and the industry's need for continued innovation. The Committee provides no less than the fiscal year 2018 funding level for ARS to support academic and Federal researchers to pursue efforts in crop protection, breeding, mechanization and other areas through USDA's Floriculture and Nursery Research Initiative.

Food Systems

The Committee recommendation includes $3,000,000 for ARS to support a Food Systems Center at a land-grant institution that addresses how local, regional and global food systems can provide nutritious and culturally appropriate food, regardless of individual life circumstances.

Foodborne Pathogens

Salmonella continues to cause serious disease in food animals and, via transmission through contaminated food products to people, remains the number one bacterial foodborne pathogen in humans. The Committee provides an additional $1,000,000 to develop non-antibiotic interventions to inhibit environmental movement of Salmonella between food animal species, and to reduce the pathogen load in food animals themselves, using Salmonella-targeted viruses called bacteriophages, as well as prebiotic and probiotic supplements.

Forage Production Systems

The Committee recommendation includes no less than the fiscal year 2018 level to develop management practices that improve the production efficiency of grazing operations in temperate pastures.

Forest Products

The Committee recognizes the important role of the forests products sector to the U.S. economy. The need to create new and improved value-added products and renewable energy from our Nation's wood supply is critical to the sustainability of the national economy. The Committee recommendation includes no less than the fiscal year 2018 level to support research on wood product quality improvement and improvement in forest products evaluation standards and valuation techniques. ARS shall conduct this research in consultation with the Forest Products Laboratory.

Fruit Fly and Exotic Pest Control

The Committee recommendation includes $1,000,000 to implement recommendations issued pursuant to Senate Report 115–131 to provide additional support and capacity to prevent the spread of fruit flies and other exotic pests to the U.S. mainland from the tropical Pacific.

Genetic Oat Research

The Committee recognizes the potential genetic oat research has to improve disease resistance (especially rusts and viruses), genetics, increase yields, and develop crop rotation systems that include oat, which will enhance the value of oats and provide benefits to producers and consumers. The Committee includes an increase of $1,000,000 to expand existing research focused on oat improvement.

Genomes to Fields

The Committee recommendation includes no less than the fiscal year 2018 level to support the Germplasm Enhancement of Maize [GEM] project to complement the existing USDA maize germplasm programs and support the emerging large-scale public sector effort to investigate the interaction of maize genome variation and environments, known as the Genomes to Fields project.

High Performance Computing Support

The Committee provides no less than the fiscal year 2018 level to support high performance computing capability to address scientific needs and directs ARS to collaborate with appropriate partners with the technical capacity and scientific synergy to provide cost-effective high performance computing support.

Hops Research

The Committee recommends no less than the fiscal year 2018 level to support hops research.

Human Nutrition Research

The Committee remains concerned about the high rates of obesity in this country, and believes that research into human nutrition is important to help prevent childhood obesity and the medical issues obesity brings. The Committee recommendation includes an increase of $1,400,000 to expand research regarding the growth, health promotion, diet, immune function, and disease prevention of the developing child.

Industrial Hemp Germplasm

The Committee recognizes the increasing demand for industrial hemp for a variety of uses, and its growing importance as a crop for U.S. farmers. When the nation's industrial hemp germplasm was destroyed in the 1980s, researchers lost access to publicly available germplasm for plant breeding purposes. The Committee directs ARS to establish and maintain a hemp germplasm repository at the Plant Genetics Research Unit and provides $500,000 for this purpose.

National Agricultural Library

The Committee strongly encourages the Agricultural Research Service to maintain its focus on agriculture-related legal issues within the National Agricultural Library. The Committee notes that as the agriculture sector faces increasing financial stress, there is a necessity that agriculture-related legal issues be addressed on an increasingly frequent basis. Further, agricultural-related legal issues are increasingly complex and the impact of these legal issues continues to broaden in scope. Therefore, the Committee recommendation includes no less than the fiscal year 2018 enacted level for the National Agricultural Library to support the Agricultural Law Information Partnership.

National Apple Rootstock Breeding Program

The Committee recognizes the importance of the National Apple Rootstock Breeding Program, which provides virus-free rootstock to apple growers throughout the Nation. Therefore, the Committee recommendation includes no less than the fiscal year 2018 level to support the ARS National Apple Rootstock Breeding Program.

National Bio and Agro-Defense Facility Transition

The Committee understands that the Department proposes to transfer the responsibility for operational planning and future operations of the National Bio and Agro-Defense Facility (NBAF) to USDA. With the transition of responsibilities from DHS to USDA in process, the Committee directed DHS in fiscal year 2018 to maintain the ability to readily execute a management, operations, and research support contract through the end of fiscal year 2018 if DHS or USDA determine that its utilization would expedite or enhance NBAF's ability to be fully operational by December 31, 2022. While the Committee appreciates that the joint DHS/USDA transition team is working to identify and outline transition activities to address all requirements for the timely operational stand-up of NBAF, the Committee is concerned that the full breadth of transition activities remain undetermined and have not been delivered to Congress. As such, the Committee directs USDA to submit a report to the Committee no later than 60 days after enactment of this Act detailing the transition activities, business plan, and critical milestones to reach the full operational capability of NBAF by December 31, 2022. The report shall include specific information on workforce development such as efforts to rapidly recruit and hire talented personnel, the number of positions that must be filled at each critical milestone, and requirements for each position. Further, the report should include specific details regarding USDA's potential plan to seize control and utilize current contract mechanisms under DHS, and the funding associated for such utilization, while remaining within the bounds of Federal Acquisition Regulations. Without a transition plan to provide greater fidelity, the Committee is concerned that neither DHS nor USDA will be able to complete commissioning activities and on-going planning activities at the outset of fiscal year 2019 and remain on-schedule absent a support contract that DHS anticipated executing to carry out such efforts. To ensure USDA can meet critical milestones, prevent delays, and adhere to the current schedule for NBAF, the Committee directs no less than $7,000,000 for USDA to execute a management, operations, and research support contract to expedite the hiring of a capable work-force for the Central Utility Plant of NBAF to ensure the timely completion of commissioning activities and on-going operations planning activities coordinated by DHS and USDA. Additionally, the Committee provides $10,600,000 to address one-time costs associated with the transfer of the science program from the Plum Island Animal Disease Center to NBAF and $42,000,000 to address stand-up activities and other initial costs to operate and maintain the facility.

NBAF will provide the U.S. with expanded capacity to implement a comprehensive bio-defense research program to protect against foreign animal diseases that pose the greatest threats to animal agriculture and public health. The Committee provides an additional $5,000,000 for ARS to increase research efforts on foreign animal diseases and emerging diseases with high consequence to animal and public health.

Nutrient Density Profile

The ARS is directed to update the nutrient profile and nutrient density characterization of pure maple syrup.

Nutrition Research and Aging

The Committee recognizes the critical importance of human nutrition research and its significance for preventative healthcare and degenerative and age-related diseases. More research is needed to address the needs of all Americans, with a particular focus on the elderly, the fastest growing segment of the population. Therefore, the Agricultural Research Service is directed to prioritize human nutrition research across the lifespan.

Pear Genetics and Genomics

The Committee recognizes that research into pear genetics and genomics is needed to identify genetic sources of pest resistance and to contribute to improved, size-controlling rootstocks to enhance orchard efficiency, and to otherwise improve cultivated pear research. The Committee recommendation includes no less than the fiscal year 2018 level to support research into pear genetics and genomics.

Postharvest Dairy Research

The Committee recognizes the importance of developing solutions to address agricultural postharvest inefficiencies to conserve limited resources and feed a growing population. The Committee provides an increase of $1,500,000 for research to develop postharvest technologies that decrease waste and improve resource use of protein, fat, and sugar in dairy processing.

Poultry Research

The Committee recognizes the important role of the poultry sector to the U.S. economy. The Committee provides an additional $1,000,000 to expand the research capacity for poultry production and health.

Public Health Research

The Committee strongly supports USDA research but is concerned about the use of cats in painful and terminal laboratory experiments at USDA's Animal Parasitic Disease Laboratory. The Committee appreciates USDA's responsiveness to concerns that have been raised and directs the agency to consult scientific and veterinary experts about the feasibility of implementing alternatives to the use of cats in public health research, and to develop a program to adopt out cats no longer needed in research. The Committee directs the Secretary to provide a report on its progress no later than 90 days after the enactment of this Act.

Pulse Health Initiative

The Committee supports the expansion of pulse crop research and provides an additional $1,000,000 to enhance scientific research into the health and nutritional benefits of dry peas, lentils, chickpeas and dry beans.

Rangeland Research

The Committee recognizes the demonstrated potential for cooperative partnerships to address complex sagebrush steppe ecosystem challenges in the Great Basin region. The Committee recommendation includes no less than the fiscal year 2018 level for ARS to support a regional, multi-institutional cooperative partnership to advance collaborative science-based conservation research, extension, and education to address time-sensitive and shared rangeland challenges affecting sustainable agricultural productivity, rural communities, and ecosystem health.

Research Assistance

The Committee encourages the Agricultural Research Service to provide direct, place-based assistance to 1862 Institutions in States that do not have Agricultural Research Service facilities to address the research priorities of such States, such as invasive plant species and insects that cause significant impacts to agriculture, aquaculture, and communities in such States and to assist in the development of specialty and horticultural crops to increase food security and expand marketing opportunities for small farmers. The Committee directs ARS to submit a report on the prospective options of such assistance.

Resilient Dryland Farming

The Committee recognizes the need for advancements in dryland production practices, cropping, and equipment to increase profitability, conserve the soil, enhance soil water storage, promote soil health, and decrease reliance on herbicides. The Committee provides an additional $2,000,000 to expand research focused on resilient dryland farming. Research should focus on improving yield and quality parameters; developing cropping systems capable of tolerating drought, heat, and diseases; and quantifying economic and environmental benefits from dryland crop production systems.

Roseau Cane

The Committee is concerned with the invasive species scale insect pest that is destroying Roseau cane in the Mississippi River's Delta region along the Gulf of Mexico. An estimated 225,000 acres of wetlands in the Delta have been affected with the die-off, and Roseau cane is important in maintaining a healthy marsh and preventing erosion. The Committee directs ARS to work with the Animal and Plant Health Inspection Service [APHIS] and stakeholders to develop an integrated management program for control of the Roseau cane scale insect pest infestation.

Sclerotinia

The Committee is aware of the economic importance of controlling sclerotinia, which affects sunflowers, soybeans, canola, edible beans, peas, and lentils and encourages ARS to continue both core research and cooperative projects of the National Sclerotinia Initiative.

Shellfish Research

The Committee encourages the Agricultural Research Service to increase its investment in partnerships with research institutions on research to improve shellfish survival and growth rates and to classify and preserve natural genetic variation. Therefore, the Committee includes an increase of $500,000 to support shellfish genetics research.

Soft White Wheat Falling Numbers Test

The Committee recognizes the emerging crisis surrounding wheat starch degradation, as detected by the Hagberg-Perten Falling Numbers [FN] Test. The quality loss was particularly devastating to Pacific Northwest soft white wheat producers in late 2016. The Committee recommendation includes no less than the fiscal year 2018 funding level to research the accuracy of the FN test, and better understand environmental, storage, and genetic conditions leading to this quality loss.

Small Grains Genomic Initiative

The Committee supports research on barley and wheat high throughput genomics and phenotyping and recognizes its importance in improving crop traits and developing new cultivars. The Committee recommendation includes an additional $1,000,000 to support the Small Grains Genomic Initiative.

Sorghum Genetic Database

The Committee recognizes the importance of phenotyping and genotyping that allows breeders to understand which genes are responsible for improvements in pest resistance, drought tolerance, and yield. The Committee recommends an increase of no less than the fiscal year 2018 level to further facilitate the partnership between ARS and the Department of Energy on sorghum genome mapping—particularly the creation of an easily-accessed database to house the information generated from the ongoing genetic sequencing research which will facilitate further crop development efforts, especially in combating the sugarcane aphid, a new and devastating invasive pest.

Sudden Oak Death

The European strain 1 [EU1] and the North American strain 1 [NA1] of the sudden oak death pathogen are major threats to western Douglas-fir/tanoak forests, resulting in quarantine restrictions that threaten US forests and export markets for log shipments and lily bulbs. The Committee recommendation includes an increase of $1,200,000 for research to improve understanding of the European Strain 1 and North American Strain 1 of the sudden oak death pathogen and treatment methods to inform control and management techniques in wildlands.

Sugar Beet Research

The Committee provides an additional $1,000,000 for plant disease research to improve the quality of sugar beet production.

Sugarcane Variety Development

The Committee recognizes the devastating impact wrought by invasive pests on the domestic sugarcane industry and provides an additional $500,000 to support the development of new pest and disease-resistant varieties.

Sustainable Aquaculture

The Committee notes that aquaculture is the fastest growing food production industry in the world and encourages ARS, in partnership with universities, to support rapid response research on sustainable aquaculture for coldwater and warmwater production environments, with an emphasis on workforce education.

Sustainable Water Use Research

The alluvial plain within the Lower Mississippi River Basin is one of most productive agricultural regions in the United States. The Committee remains concerned with the unsustainable use of water in the Alluvial Aquifer as a result of increasing water withdrawals and stagnant recharging. The Committee provides no less than the fiscal year 2018 level for research to improve the recharge capabilities of the Alluvial Aquifer and to develop new conservation and irrigation techniques to reduce water usage in agriculture production.

Tropical and Subtropical Research

Research on Tropical and Subtropical crops is critical as the presence of and destruction by invasive pests such as fruit flies, coffee berry borer, felted macadamia nut coccid, and plant viruses and fungal diseases increasingly threaten crop security in the Pacific and Insular Areas, and the Committee encourages ARS to support this research.

UAS Precision Agriculture Applications

The Committee provides no less than the fiscal year 2018 level to support efforts utilizing unmanned aerial systems [UAS] in crop production operations and to address the challenges associated with data capture, transfer and analysis.

U.S. Wheat and Barley Scab Initiative [USWBSI]

The Committee recognizes that fusarium head blight is a major threat to agriculture, inflicting substantial yield and quality losses throughout the U.S. The Committee supports research carried out through the USWBSI. The Committee recommendation includes no less than the fiscal year 2018 level to conduct further research on reducing the impact of fusarium head blight on wheat and barley.

Warmwater Aquaculture

The Committee provides an additional $1,600,000 to facilitate the advancement of technologies that improve the efficiency, profitability and sustainability of warmwater aquaculture production.

Wheat and Sorghum Research

The Committee recognizes the potential impact heat and drought can have on the yield and quality of wheat and sorghum and the need for new cultivars to adapt to changing climatic conditions. In addition, sorghum crops have been particularly hit hard by the invasive sugarcane aphid and new resistant cultivars are needed. The Committee provides an additional $1,000,000 for research to improve the productivity and quality of wheat and sorghum during uncertain growing seasons resulting from extended droughts and increased temperatures. Within this increase, funding is included to initiate gene flow research to advance the durability and sustainability of fitness traits in sorghum.

Buildings and Facilities

Appropriations, 2018	$140,600,000
Budget estimate, 2019 Committee recommendation	

Committee Recommendations

The Committee does not recommend an appropriation for Agricultural Research Service, Buildings and Facilities.

National Institute of Food and Agriculture

Section 7511(f)(2) of the Food, Conservation, and Energy Act of 2008 amends the Department of Agriculture Reorganization Act of 1994 (7 U.S.C. 6971) by establishing an agency to be known as the National Institute of Food and Agriculture [NIFA]. The Secretary transferred to the Director of NIFA, effective October 1, 2009, all authorities administered by the Administrator of the Cooperative State, Research, Education and Extension Service. The mission is to work with university partners and customers to advance research, extension, and higher education in the food and agricultural sciences and related environmental and human sciences to benefit people, communities, and the Nation.

Research and Education Activities

Appropriations, 2018	$887,171,000
Budget estimate, 2019	794,479,000
Committee recommendation	898,535,000

Research and Education programs administered by NIFA are USDA's principal entrée to the university system of the United States for the purpose of conducting agricultural research and education programs as authorized by the Hatch Act of 1887, as amended (7 U.S.C. 361a–361i); the McIntire-Stennis Cooperative Forestry Act of 1962, as amended (16 U.S.C. 582a et seq.); the Competitive, Special, and Facilities Research Grant Act, as amended (7 U.S.C. 450i); the National Agricultural, Research, Extension, and Teaching Policy Act of 1977, as amended (7 U.S.C. 3101 et seq.); the Equity in Educational Land-Grant Status Act of 1994 (7 U.S.C. 301 note); the Agricultural Research, Extension and Education Reform Act of 1998 (Public Law 105–185), as amended; the Food, Agriculture, Conservation and Trade Act of 1990 (Public Law 101–624); the Farm Security and Rural Investment Act of 2002 (Public Law 107–171); and the Food, Conservation and Energy Act of 2008 (Public Law 110–246). Through these authorities, USDA participates with State and other cooperators to encourage and assist the State institutions in the conduct of agricultural research and education through the State Agricultural Experiment Stations of the 50 States and the territories; by approved Schools of Forestry; the 1890 Land-Grant Institutions and Tuskegee University and West Virginia State University; 1994 Land-Grant Institutions; by Colleges of Veterinary Medicine; and other eligible institutions. The appropriated funds provide Federal support for research and education programs at these institutions.

The research and education programs participate in a nationwide system of agricultural research program planning and coordination among the State institutions, U.S. Department of Agriculture, and the agricultural industry of America.

Committee Recommendations

The Committee recommends an appropriation of $898,535,000 for research and education activities of the National Institute of Food and Agriculture.

The following table summarizes the Committee's recommendations for research and education activities:

National Institute of Food and Agriculture: Research and Education Activities

[Dollars in thousands]		
Program/Activity	**Authorization**	**Committee recommendation**
Hatch Act	7 U.S.C. 361a–i	243,701
McIntire-Stennis Cooperative Forestry Act	16 U.S.C. 582a through a–7	36,000
Research at 1890 Institutions (Evans-Allen Program)	7 U.S.C. 3222	54,185
Payments to the 1994 Institutions	534(a)(1) of Public Law 103–382	3,439
Education Grants for 1890 Institutions	7 U.S.C. 3152(b)	19,336
Education Grants for Hispanic-Serving Institutions	7 U.S.C. 3241	9,219
Education Grants for Alaska Native and Native Hawaiian-Serving Institutions.	7 U.S.C. 3156	3,194
Research Grants for 1994 Institutions	536 of Public Law 103–382	3,801
Capacity Building for Non Land-Grant Colleges of Agriculture	7 U.S.C. 3319i	5,000
Resident Instruction and Distance Education Grants for Insular Areas.	7 U.S.C. 3362 and 3363	2,000
Agriculture and Food Research Initiative	7 U.S.C. 450i(b)	405,000
Veterinary Medicine Loan Repayment	7 U.S.C. 3151a	8,000
Veterinary Services Grant Program	7 U.S.C. 3151b	3,000
Continuing Animal Health and Disease Research Program	7 U.S.C. 3195	4,000
Supplemental and Alternative Crops	7 U.S.C. 3319d	1,000
Multicultural Scholars, Graduate Fellowship and Institutions Challenge Grants.	7 U.S.C. 3152(b)	9,000
Secondary and 2-year Post-Secondary Education	7 U.S.C. 3152(j)	900
Aquaculture Centers	7 U.S.C. 3322	5,000
Sustainable Agriculture Research and Education	7 U.S.C. 5811, 5812, 5831, and 5832.	37,000

(Continued).

[Dollars in thousands]		
Program/Activity	**Authorization**	**Committee recommendation**
Farm Business Management	7 U.S.C. 5925f	2,000
Sun Grant Program	7 U.S.C. 8114.	3,000
Minor Crop Pest Management (IR–4)	7 U.S.C. 450i(c)	11,913
Alfalfa Forage and Research Program	7 U.S.C. 5925	3,000
Special Research Grants:7 U.S.C. 450i(c):		
Global Change/UV Monitoring		1,405
Potato Research		2,750
Aquaculture Research		2,000
Total, Special Research Grants		6,155
Necessary Expenses of Research and Education Activities:		
Grants Management System		7,830
Federal Administration—Other Necessary Expenses for Research and Education Activities.		11,862
Total, Necessary Expenses		19,692
Total, Research and Education Activities	.	898,535

Agricultural Research Enhancement Awards

The Committee remains determined to see that quality research and enhanced human resources development in the agricultural and related sciences be a nationwide commitment. Therefore, the Committee continues its direction that not less than 15 percent of the competitive research grant funds be used for USDA's agricultural research enhancement awards program, including USDA–EPSCoR.

Agriculture and Food Research Initiative

The Committee recommendation includes $405,000,000 for the Agriculture and Food Research Initiative [AFRI].

Section 7406 of the Food, Conservation, and Energy Act of 2008 specifies priority areas within the Agriculture and Food Research Initiative [AFRI], including an emphasis on conventional (classical) plant and animal breeding. The Committee notes the importance of this requirement to provide farmers nationwide with greater access to cultivars that are locally and regionally adapted to their soils, climates and farming systems. Because of the agency's lack of progress in prioritizing this effort, the Committee directs the agency to make regionally adapted, publicly held cultivar development a distinct funding priority within AFRI for fiscal year 2019, and directs the agency to take steps to improve its tracking of public cultivar projects within AFRI and report its progress in meeting this goal.

Agriculture Technology

The Committee encourages USDA to support research and development of agricultural robotics, particularly to increase yields in vertically stacked farming production.

Alfalfa and Forage Research

The Committee notes that research into alfalfa and forage holds the potential to increase alfalfa and forage yields, increase milk production, and improve forage genetics to increase biomass for the production of cellulosic ethanol. The Committee recommendation includes $3,000,000 to support research into the improvement of yields, water conservation, creation of new uses of alfalfa and forage for bioenergy, and the development of new storage and harvest systems.

Algae Applications in Agriculture Research

The Committee encourages NIFA to support research on algae and algae application in agriculture.

Aquaculture Disease Research

The Committee encourages USDA to support aquaculture disease and vaccine research, including research on coldwater aquaculture vaccines. There is currently no national facility for pathogen testing. Research into finfish vaccines and pathogens has the potential to accelerate the growth of sustainable U.S. aquaculture, reduce the trade deficit attributable to imported seafood, and reduce the pressure on overfished species.

Aquaculture Research

The Committee recognizes the importance of the domestic aquaculture industry to the U.S. economy. The Committee recommendation includes $2,000,000 for aqua-culture research to address issues related to genetics, disease, systems, and economics.

Brucellosis Research

Federal and State animal health officials have made eradicating livestock disease with significant reservoirs a national animal health priority. This need was reflected in the Agricultural Act of 2014, which made the research and development of surveillance methods, vaccines, vaccination delivery systems or diagnostics tests a priority research area under the Competitive, Special, and Facilities Research Grant Act particularly for bovine brucellosis and bovine tuberculosis. The Committee recognizes the need for this research and encourages the agency to make competitive grants available to study improved management tools for zoonotic livestock diseases with significant wildlife reservoirs.

Cereal Crop Research

Research on cereal crops has historically been conducted by USDA and public universities, and the Committee recognizes the importance of continuing investment in cereal crop research. The Committee strongly encourages USDA to provide funding for cereal crop research in the areas of genetic and genomic research, plant pest research, and improved production systems.

Childhood Obesity

The Committee encourages USDA to support innovative efforts to address the unique challenges faced in addressing obesity among children and youth in urban, minority low-income populations through a combination of family education and clinical studies.

Citrus Disease Research Program

The 2014 Farm Bill established the Emergency Citrus Disease Research and Extension Program, which is intended to discover and develop tools for early detection, control, and eradication of diseases and pests that threaten domestic citrus production and processing, and provided $25,000,000 per year in mandatory funding for the program through the Specialty Crop Research Initiative. The Committee believes research projects funded under this authority should be prioritized based on the critical threat of citrus greening and encourages NIFA, to the maximum extent practicable, to follow the recommendations of the National Agricultural Research, Extension, and Education Advisory Board's citrus disease subcommittee and to collaborate with the HLB–MAC group.

Community College Centers of Excellence in Agribusiness Workforce Training

The Committee encourages NIFA to designate Community College Centers of Excellence in Agribusiness Workforce Training, to include a limited number of 2-year community and technical colleges with a demonstrated capability to provide training and education for Agribusiness. The Centers of Excellence will seek to develop model programs in Agribusiness and promote economic development.

Countering Seafood Fraud

The Committee remains concerned about countering economic fraud and improving food safety of the U.S. food supply. The Committee is concerned that adequate technology is not yet available to provide for appropriate sampling of the food supply. The Committee believes NIFA should conduct research to develop technologies that will provide rapid, portable and facile screening of fish species at port sites, wholesale, and retail centers.

Diversification in Agriculture

The Committee recognizes the rapid evolution of U.S. agriculture, including the diversification of practices, markets, and technologies as farms transition to one generation from another, and encourages NIFA to prioritize investments that deliver hands-on technical education in diversified agriculture and food systems and support technical colleges seeking to establish beginning farmer programs serving diversified agriculture, and aid in supporting farm viability.

Dual Use/Dual Benefit

The Dual Purpose with Dual Benefit: Research in Biomedicine and Agriculture Using Agriculturally Important Domestic Species is an interagency partnership grants program funded by the National Institute of Child Health and Human Development [NICHD] and USDA. The Committee strongly urges continuation of this partnership because it sponsors use of farm animals as dual purpose models to better understand developmental origins of disease, fat regulation and obesity, stem cell biology, assisted reproductive technologies, and infectious disease, which directly benefits both agriculture and biomedicine. This program also strengthens ties between human medicine, veterinary medicine, and animal sciences, which is key to success of the One Health Initiative.

Food Safety

The Committee recommends that NIFA prioritize research on funding for new food safety technologies relating to the Nation's meat supply that helps researchers, producers, and manufacturers.

Food Safety and Defense Technology

The Committee is concerned that insufficient progress is being made in the development of detection technology in the food safety sector. The ability to rapidly, accurately, and cost effectively detect pathogens or contaminants throughout the food supply chain is critical to protecting the United States from food-borne illnesses and malicious acts. As such, the Committee encourages NIFA to increase research of novel biodetection technologies and the implementation of mobile bio-detection platforms in real-world conditions. The Department should consider technologies currently in use or under development in other fields, such as medicine or homeland security, to determine whether the technology can meet the needs in either high volume food production or mobile food defense monitoring.

Foodborne Illness Prevention

The Committee understands the significant threats to public health and to the economic viability of communities impacted by foodborne illness, and believes that coordinated and targeted resources are needed to understand the risks and to develop effective strategies for control. The Committee encourages NIFA, in coordination with the FDA, to establish a Center of Excellence for Foodborne Illness to coordinate a research program to reduce the risk of Listeria monocytogenes.

Genomes to Phenomes

The Committee is supportive of the multi-university crop research initiative known as Genomes to Phenomes, and encourages NIFA to support the development of tools and datasets that can be used across multiple crop species to improve the output and efficiency of agriculture.

Lowbush Blueberries

The Committee directs NIFA to work with research institutions to develop and refine predictive models and monitoring technologies for native and invasive pests for incorporation into integrated pest management programs for naturally seeded, native berry crops to increase the margin of food safety and product quality.

Multi-Trophic Aquaculture Research

Nearly half the seafood consumed across the world is the result of aquaculture, and the aquaculture industry is a critical and growing part of the U.S. economy. However, less than 1 percent of worldwide production comes from U.S. producers. The Committee is concerned that inefficient production technologies hinder the ability of the domestic aquaculture industry to compete on a global scale. The Committee supports development and demonstration of an integrated aqua-culture system that would contain at one site a highly competitive and sustainable system having a low environmental footprint and be primarily self-contained. The Committee supports the development of a "Beta" model that would focus on developing, building, operating, demonstrating and teaching around this intensified, integrated, bio-secure production technology for feed, fish-plant and energy products.

Oak Mites

The Committee directs NIFA to study the recent infestation of oak mites and focus on suppression and eradication possibilities.

Organic Research

USDA's National Organic Standards Board [NOSB] has identified key organic research priorities, many of which would help to address challenges that have limited the growth in organic production in this country. The Committee encourages NIFA to give strong consideration to the NOSB organic research priorities when crafting the fiscal year 2019 RFAs for AFRI and the Organic Transition Program. Given the growing demand for organic products, the Committee also encourages USDA to increase the number of organic research projects funded under AFRI and Specialty Crop Research Initiative [SCRI].

Protein Functionality

The Committee encourages USDA to support research projects that characterize protein from crop plants such as chickpeas, sorghum, lentils, fava beans, lupin, rice, oats, mushrooms, and water lentils to assess their suitability for use in food products. The Committee is particularly interested in research projects involving plants that can be easily cultivated in the U.S. and that are sustainably grown and produced (in terms of factors such as water usage or fertilizer and pesticide requirements).

Public Plant and Animal Breeding

The Committee is concerned about the decline in public plant and animal breeding programs at our nation's land-grant institutions over the last 25 years, and encourages LGUs to take steps to foster the next generation of public plant and animal breeders by placing a higher priority on the development of publicly available, regionally adapted cultivars and breeds. For all regions of our nation to optimize their productive capacity in an environmentally sustainable manner, it is critical that the farmers of the region have access to the most up-to-date cultivars and breeds to meet ever-changing conditions.

Regional Research Priorities

The Committee encourages NIFA to consider providing funding within AFRI to assist with State and regional research priorities, with USDA oversight and review.

Seafood

The Committee encourages USDA, in partnership with universities with established domestic shrimp farming programs, to support the development of a domestic industry that will help ensure the safety and quality of the Nation's seafood supply, promote environmentally sustainable aquaculture, create new opportunities for U.S. agriculture, and forge new markets for U.S. grain and oil-seed products and technology services.

Small Fruits Research

The Committee encourages USDA to support research to promote sustainable production of berry and grape crops with the goal of reducing pesticide use and improving quality and yield. Additionally, the Committee encourages USDA to support research to improve the ability to forecast pest and disease spread, and implement precision management strategies.

Specialty Crop Research Initiative

The Committee emphasizes the important role of the Specialty Crop Research Initiative in addressing the critical needs of the specialty crop industry through research and extension activities, and encourages NIFA to prioritize proposals for and enhance its overall commitment to identifying and addressing threats to pollinators from pests and diseases.

Supplemental and Alternative Crops

The Committee recognizes the importance of nationally coordinated, regionally managed canola research and extension programs. The Committee encourages the Secretary to continue to seek input from stakeholders and to give priority consideration to proposals in the peer review process that address research needs in production areas with the greatest potential to expand as well as those where canola production is established and needs to be maintained.

Sustainable Agriculture Research and Education

The Committee is strongly supportive of the SARE program and directs USDA to ensure that research, education and extension activities carried out within SARE remain intact.

The Committee believes that it is important for NIFA to evaluate the performance of each of its regional SARE Host Institutions on a regular basis; however the Committee is concerned that the recent change in practice for NIFA to broadly solicit competitive Host Institution proposals at least every 5 years may interfere with the ability of the regional Host to retain qualified staff, and to establish a stable operating base, hindering rather than enhancing SARE program delivery. Not later than 90 days after the date of enactment of this Act, the Secretary shall report to the Committees on Appropriations on whether frequent open competitions for SARE Host Institutions optimizes the delivery of the SARE program, as compared to previous performance reviews and competitive solicitations that were performed every 10 years.

Unmanned Aerial Systems in Agriculture

The Committee encourages USDA to support regional institutes focused on the development of Unmanned Aerial Systems [UAS] and fostering new innovations in agriculture and cybersecurity. UAS is a tool to obtain data in a wide variety of application areas including energy, agriculture, power infrastructure, and transportation, all critical to rural States. The Committee encourages NIFA to support the research, development, and expansion of the use of UAS and high performance computing.

Veterinary Corps

Veterinarians fulfilling the terms of a contract under USDA's Veterinary Medicine Loan Repayment Program, authorized by the National Veterinary Medical Services Act, shall be members of the National Veterinary Medical Services Corps and members who have fulfilled the terms of their contract shall be alumni of the Corps.

Native American Institutions Endowment Fund

Appropriations, 2018	$11,880,000
Budget estimate, 2019	11,857,000
Committee recommendation	11,880,000

The Native American Institutions Endowment Fund authorized by Public Law 103–382, the Equity in Educational Land-Grant Status Act, provides an endowment for the 1994 land-grant institutions (34 tribally controlled colleges). This program will enhance educational opportunity for Native Americans by building educational capacity at these institutions in the areas of student recruitment and retention, curricula development, faculty preparation, instruction delivery systems, and scientific instrumentation for teaching. Income funds are also available for facility renovation, repair, construction, and maintenance. On the termination of each fiscal year, the Secretary shall withdraw the income from the endowment fund for the fiscal year, and after making adjustments for the cost of administering the endowment fund, distribute the adjusted income as follows: 60 percent of the adjusted income from these funds shall be distributed among the 1994 land-grant institutions on a pro rata basis, the proportionate share being based on the Indian student count; and 40 percent of the adjusted income shall be distributed in equal shares to the 1994 land-grant institutions.

Committee Recommendations

The Committee recommends an appropriation of $11,880,000 for the Native American Institutions Endowment Fund.

Extension Activities

Appropriations, 2018	$483,626,000
Budget estimate, 2019	450,185,000
Committee recommendation	486,692,000

Cooperative extension work was established by the Smith-Lever Act of May 8, 1914, as amended. The Department of Agriculture is authorized to provide, through the land-grant colleges, cooperative extension work that consists of the development of practical applications of research knowledge and the giving of instruction and practical demonstrations of existing or improved practices or technologies in agriculture and related subjects, and to encourage the application of such information by demonstrations, publications, through 4–H clubs, and other means to persons not in attendance or resident at the colleges.

To fulfill the requirements of the Smith-Lever Act, State and county extension offices in each State, the District of Columbia, Puerto Rico, the Virgin Islands, Guam, American Samoa, the Northern Marianas, and Micronesia conduct educational programs to improve American agriculture and strengthen the Nation's families and communities.

Committee Recommendations

The Committee recommends an appropriation of $486,692,000 for extension activities of the National Institute of Food and Agriculture.

The following table summarizes the Committee's recommendations for extension activities:

Peer-Reviewed Accreditation

The Committee notes APHIS's collaboration with accrediting organizations in the establishment and operation of the Zoo and Aquarium Hazards Preparedness, Response and Recovery Fusion Center. The Center's role in facilitating and enhancing collaboration between the emergency management community and the greater managed wildlife community will assist in emergency preparedness and response for a wide range of circumstances including natural disasters and animal disease outbreaks. This effort is commended and is expected and encouraged to continue.

Regional Biosecurity Plan for Micronesia and Hawaii

The Secretary is directed to submit a report to this Committee on USDA activities to implement the Regional Biosecurity Plan for Micronesia and Hawaii [RBP]. The report shall include an update on agencies' activities to date to implement the RBP, and agencies' planned activities for further implementation.

Roseau Cane

The Committee is concerned with the invasive species scale insect pest that is destroying Roseau cane in the Mississippi River's Delta region along the Gulf of Mexico. An estimated 225,000 acres of wetlands in the Delta have been affected with the die-off, and Roseau cane is important in maintaining a healthy marsh and preventing erosion. The Committee directs APHIS to work with the Agricultural Research Service [ARS] and stakeholders and provides an additional $500,000 to develop an integrated management program for control of the Roseau cane scale insect pest infestation.

Sudden Oak Death

The European strain 1 [EU1] and the North American strain 1 [NA1] of the sudden oak death pathogen are major threats to western Douglas-fir/tanoak forests, resulting in quarantine restrictions that threaten U.S. forests and export markets for log shipments and lily bulbs. The

National Institute of Food and Agriculture: Extension Activities

[Dollars in thousands]		
Program/Activity	**Authorization**	**Committee recommendation**
Smith-Lever Act, Section 3(b) and (c) and Cooperative Extension	7 U.S.C. 343(b) and (c) and 208(c) of Public Law 93–471.	300,000
Extension Services at 1890 Institutions	7 U.S.C. 3221	45,620
Extension Services at 1994 Institutions	7 U.S.C. 343(b)(3)	6,446
Facility Improvements at 1890 Institutions	7 U.S.C. 3222b	19,730
Renewable Resources Extension Act.	16 U.S.C. 1671 et seq.	4,060
Rural Health and Safety Education Programs	7 U.S.C. 2662(i)	3,000
Food and Animal Residue Avoidance Database Program	7 U.S.C. 7642	2,500
Women and Minorities in STEM Fields	7 U.S.C. 5925	400
Food Safety Outreach Program	7 U.S.C. 7625	8,000
Food and Agriculture Service Learning	7 U.S.C. 7633	1,000
Smith-LeverAct, Section 3(d):		
Food and Nutrition Education	7 U.S.C. 343(d)	70,000
Farm Safety and Youth Farm Safety Education Programs	7 U.S.C. 343(d)	4,610
New Technologies for Agricultural Extension	7 U.S.C. 343(d)	1,550
Children, Youth, and Families at Risk	7 U.S.C. 343(d)	8,395
Federally Recognized Tribes Extension Program	7 U.S.C. 343(d)	3,039
Total, Section 3(d)		87,594
Necessary Expenses of Research and Education Activities:		
Agriculture in the K–12 Classroom		552
Federal Administration—Other Necessary Expenses for Research and Education Activities.		7,790
Total, Necessary Expenses		8,342
Total, Extension Activities		486,692

Farmer Stress Assistance Network

The Committee is deeply concerned by the high rate of suicides among agriculture workers, which is the highest overall suicide rate among all occupations according to the Department of Health and Human Services. The Committee provides $2,000,000 in a general provision for a pilot program to provide competitive grants to State departments of agriculture, State cooperative extension services, and nonprofit organizations to carry out programs to address farmer stress and suicide. The Secretary is directed to submit a report on implementation within 60 days of enactment.

Food and Nutrition Education

The Committee recognizes the importance of the Expanded Food Nutrition Education Program [EFNEP] and encourages the Secretary to support a special pilot expansion of EFNEP to provide for an evaluation of improved food resource management and diet quality in populations not now served, including the elderly, households living below 185 percent of the poverty level, and low-income households with children of any age.

Minority Outreach

The Committee is concerned that extension service resources do not reach minority, socially disadvantaged, and tribal communities in proportion to their participation in the agricultural sector. All institutions that receive extension activity funding should seek to ensure that an equitable percentage of their overall extension work reaches minority, socially disadvantaged, and tribal communities. The Committee directs NIFA to evaluate distribution of extension resources to these three populations and report to the Committee no later than 90 days after enactment of this Act.

Integrated Activities

Appropriations, 2018	$37,000,000
Budget estimate, 2019	13,037,000
Committee recommendation	38,000,000

Section 406, as amended, of the Agricultural Research, Extension, and Education Reform Act of 1998 authorizes an integrated research, education, and extension competitive grants program.

Committee Recommendations

The Committee recommends an appropriation of $38,000,000 for integrated activities of the National Institute of Food and Agriculture.

The following table summarizes the Committee's recommendations for integrated activities:

National Institute of Food and Agriculture: Integrated Activities

[Dollars in thousands]		
Program/Activity	**Authorization**	**Committee recommendation**
Methyl Bromide Transition Program	7 U.S.C. 7626	2,000
Organic Transition Program	7 U.S.C. 7626	6,000
Regional Rural Development Centers	7 U.S.C. 450i(c)	2,000
Food and Agriculture Defense Initiative	7 U.S.C. 3351	8,000
Crop Protection/Pest Management	7 U.S.C. 343(d)	20,000
Total, Integrated Activities		38,000

Food and Agriculture Defense Initiative

The Committee supports the important work being done through the publicly funded diagnostic laboratory network and encourages NIFA to prioritize funding to strengthen animal health diagnostic laboratories, taking into consideration the degree to which the capacity for surveillance, monitoring, response, and capacity is enhanced; the concentration of human and animal populations that are directly at risk; trade, tourism, and cultural considerations; geography, ecology, and climate; evidence of active collaboration with, and support of, the State animal health officials;

those States with highest risk for the introduction of foreign and emerging pests and diseases; and evidence of stakeholder support and engagement.

Potato Research

To minimize the application of pesticides and to maximize the yield and quality of harvested potatoes, the Committee directs the Secretary to support pest management programs in potato growing States. Such programs help scientists track potential pest outbreaks and provide growers and industry professionals with current information on specific and timely treatments. Additionally, the programs help identify serious diseases, such as late blight disease, in their early stages, allowing for preventive measures to be put into place quickly to avoid crop losses.

Office of the Under Secretary for Marketing and Regulatory Programs

Appropriations, 2018	$901,000
Budget estimate, 2019	800,000
Committee recommendation	901,000

The Office of the Under Secretary for Marketing and Regulatory Programs provides direction and coordination in carrying out laws with respect to the Department's marketing, grading, and standardization activities related to grain; competitive marketing practices of livestock, marketing orders, and various programs; veterinary services; and plant protection and quarantine. The Office has oversight and management responsibilities for the Animal and Plant Health Inspection Service; Agricultural Marketing Service; and Grain Inspection, Packers and Stockyards Administration.

Committee Recommendations

The Committee recommends an appropriation of $901,000 for the Office of the Under Secretary for Marketing and Regulatory Programs.

Animal and Plant Health Inspection Service

Salaries and Expenses

Appropriations, 2018	$981,893,000
Budget estimate, 2019	739,151,000
Committee recommendation	1,000,493,000

The Secretary of Agriculture established the Animal and Plant Health Inspection Service [APHIS] on April 2, 1972, under the authority of reorganization plan No. 2 of 1953, and other authorities. The major objectives of APHIS are to protect the animal and plant resources of the Nation from diseases and pests. These objectives are carried out under the major areas of activity, as follows:

Safeguarding and Emergency Preparedness/Response

The agency monitors plant and animal health worldwide, and sets import polices to prevent the introduction of foreign plant and animal pests and diseases. Domestically, the agency works cooperatively to conduct plant and animal health monitoring programs, pursue eradication, or limit the spread of the threat. The agency also conducts diagnostic laboratory activities that support disease prevention, detection, control, and eradication programs. In addition, the agency protects agriculture from detrimental animal predators, and through its regulatory structure helps advance genetic research while protecting against the release of harmful organisms.

Safe Trade and International Technical Assistance

The agency helps resolve technical trade issues to ensure the smooth and safe movement of agricultural commodities into and out of the United States. The agency negotiates animal and plant health certification requirements and assists U.S. exporters meet foreign regulatory demands. In addition, the agency assists developing countries in improving their safeguarding systems, to protect the United States from emerging plant and animal pests and diseases.

Animal Care

The agency conducts regulatory activities that ensure the humane care and treatment of animals and horses as the Animal Welfare and Horse Protection Acts require. These activities include inspection of certain establishments that handle animals intended for research, exhibition, and as pets, and monitoring certain horse shows.

Committee Recommendations

The Committee recommends an appropriation of $1,000,493,000 for salaries and expenses of the Animal and Plant Health Inspection Service.

The following table reflects the Committee's specific recommendations for the Animal and Plant Health Inspection Service:

Animal and Plant Health Inspection Service

[In thousands of dollars]			
	Fiscal year 2018 enacted	**Fiscal year 2019 budget request**	**Committee recommendation**
Safeguarding and International Technical Assistance:			
Animal Health Technical Services	37,857	30,272	37,857
Aquatic Animal Health	2,253		2,253
Avian Health	62,840	33,881	62,840
Cattle Health	96,500	86,326	96,500
Equine, Cervid and Small Ruminant Health	20,000	16,500	20,000
National Veterinary Stockpile	5,725	3,965	5,725
Swine Health	24,800	19,753	24,800
Veterinary Biologics	16,417	16,386	16,417
Veterinary Diagnostics	39,540	42,030	50,140
Zoonotic Disease Management	16,523	15,775	16,523
Subtotal, Animal Health	322,455	264,888	333,055

[In thousands of dollars]			
	Fiscal year 2018 enacted	**Fiscal year 2019 budget request**	**Committee recommendation**
Agricultural Quarantine Inspection (Appropriated)	31,330		32,330
Cotton Pests	11,520	7,000	11,520
Field Crop & Rangeland Ecosystems Pests	9,326	7,809	11,826
Pest Detection	27,446	22,394	27,446
Plant Protection Methods Development	20,686	15,647	20,686
Specialty Crop Pests	178,170	139,500	178,170
Tree & Wood Pests	56,000	25,000	60,000
Subtotal, Plant Health	334,478	217,350	341,978
Wildlife Damage Management	108,376	46,331	108,376
Wildlife Services Methods Development	18,856	18,820	18,856
Subtotal, Wildlife Services	127,232	65,151	127,232
Animal & Plant Health Regulatory Enforcement	16,224	16,193	16,224
Biotechnology Regulatory Services	18,875	18,839	18,875
Subtotal, Regulatory Services	35,099	35,032	35,099
Contingency Fund	470	469	470
Emergency Preparedness & Response	40,966	40,688	41,466
Subtotal, Emergency Management	41,436	41,157	41,936
Subtotal, Safeguarding and Emergency Preparedness/Response	860,700	623,578	879,300
Safe Trade and International Technical Assistance:			
Agriculture Import/Export	15,599	15,070	15,599
Overseas Technical & Trade Operations	22,115	22,072	22,115
Subtotal, Safe Trade	37,714	37,142	37,714

(Continued)

[In thousands of dollars]			
	Fiscal year 2018 enacted	**Fiscal year 2019 budget request**	**Committee recommendation**
Animal Welfare:			
Animal Welfare	30,810	28,356	30,810
Horse Protection	705	696	705
Subtotal, Animal Welfare	31,515	29,052	31,515
Agency Management:			
APHIS Information Technology Infrastructure	4,251	4,243	4,251
Physical/Operational Security	5,146	5,136	5,146
Rent and DHS security payments	42,567	40,000	42,567
Subtotal, Agency Management	51,964	49,379	51,964
Total, Direct Appropriation	981,893	739,151	1,000,493

Agricultural Quarantine Inspection

The Committee recognizes that prevention of infestations of pests and diseases is much more cost effective than subsequent control or eradication. This is an important Federal responsibility and the Committee provides $32,330,000 for the agricultural quarantine inspections [AQI] function, including pre-departure and interline inspections.

The Committee is concerned that new fee regulations under the AQI program may not be equitable to small commercial aircraft. The Committee looks forward to seeing the report on AQI user fees as required by Public Law 115–141.

The Committee notes that assessing AQI treatment monitoring fees on a per-enclosure basis imposes disproportionate impacts on industry and user groups at certain key ports of entry, including ports along the southeast United States. USDA is encouraged to continue conducting a study that specifically outlines the actual costs of treatments, examines the disproportionate impact the fee has on airports and seaports in different regions of the U.S., and evaluates alternative and equitable funding mechanisms. Such report should also incorporate due consideration of the recommendations of the Treatment Fee Working Group's September 27, 2016 "Report to APHIS." USDA shall brief the Committee on the status of such study and other efforts to ensure equitable collection of revenues for vital AQI treatment monitoring efforts.

Animal Care

The Committee is concerned about dog breeders selling dogs, including dogs sold over the Internet and through other outlets, who refuse to obtain a license and comply with the Animal Welfare Act's humane care and treatment requirements and urges the Department to focus outreach, compliance assistance, and enforcement resources to these actors.

Avian Influenza

The Committee recognizes the extreme economic hardship posed to gamebird and egg farmers when flocks are determined to be infected by high and low pathogenic avian influenza, and acknowledges the severe limitations on controlled marketing available to producers of live game birds, as well as the income loss from egg production. The Committee encourages APHIS to provide full indemnity coverage for gamebird and egg operations and cease attempts to limit coverage.

Bee Pests

The Committee remains concerned with declining bee populations and the tragic implications for pollination of U.S. agriculture. The Committee directs the agency to continue priority work with other Federal and State agencies and the public to manage, suppress, and eradicate varroa mites, small hive beetles, and other pests and diseases contributing to colony collapse disorder.

Cattle Fever Ticks

The Committee appreciates the commitment by APHIS, including recent additional funding, to respond to the most recent outbreak of cattle fever ticks. The Committee encourages the agency to maintain this focus and provide adequate funding for all activities under the Cattle Fever Tick

Eradication Program [CFTEP]; heighten efforts to coordinate the response with the Department of Interior on national wildlife refuges; and provide sufficient funding for research and scientific tools to be developed that concentrate on the following: new systematic cattle fever tick treatment products with longer treatment intervals for cattle; new cattle fever tick treatment products for wildlife, especially nilgai antelope; and new or improved cattle fever tick preventative therapies, such as vaccines, for both cattle and wildlife hosts.

Chronic Wasting Disease

The Committee provides no less than $3,500,000 for cervid health activities. Within the funds provided, APHIS should give consideration to indemnity payments if warranted.

Citrus Health Response Program [CHRP]

CHRP is a national effort to maintain a viable citrus industry within the United States, maintain producer's continued access to export markets, and safeguard citrus producing states against a variety of invasive pests and diseases. These funds are designed to partner with state departments of agriculture and industry groups to address the challenges of citrus pests and diseases. In addition to the funds provided in this account, the Committee encourages APHIS to utilize the funds available in the Plant Pest and Disease Management and Disaster Prevention Programs account to the greatest extent possible in an attempt to sustain the economic viability of the citrus industry.

Coconut Rhinoceros Beetle

The Secretary is directed to continue efforts to control and eradicate the coconut rhinoceros beetle in the Pacific to prevent the spread of the beetle to the U.S. mainland, and may, if appropriate, provide additional surge funding to match Department of Defense spending for this purpose.

Cogongrass Management and Control

The Committee is concerned about the rapid spread of cogongrass and its impact on forest productivity, wildlife habitat and private landowners. The Committee provides $2,000,000 for APHIS to partner with state departments of agriculture and forestry commissions in states considered to be the epicenter of infestations, to assist with the control and treatment of cogongrass in order to slow the advancing front of this invasive plant-pest species.

Huanglongbing Emergency Response

The Committee maintains the increased funding levels for Huanglongbing Emergency Response within the Specialty Crop Pests line item included in fiscal year 2018. The Committee encourages APHIS to allocate sufficient resources in order to continue vital management, control, and associated activities to address citrus greening. The disease, for which there is no cure, has caused a reduction in citrus production by over 60 percent since 2007 in Florida alone. All citrus producing counties in Texas are under quarantine, and California has detected the disease in some backyard trees in the Los Angeles basin. The spread of this disease has called the domestic citrus industry's future into question, costing thousands of jobs and millions in lost revenue and increased production costs per acre. In addition, the agency is encouraged to support priorities and strategies identified by the Huanglongbing Multi-Agency Coordination [HLB–MAC] Group which will benefit the citrus industry. The agency should appropriately allocate resources based on critical need and maximum effect to the citrus industry. The Committee maintains the fiscal year 2018 funding level for citrus health to support priorities and strategies identified by the HLB–MAC group. The MAC is focused on short-term solutions to help the citrus industry, and the cooperative nature of Federal, state, and industry representatives in this group is expected to result in the development of tools and techniques to address this devastating disease. Helping growers explore new possible solutions, the MAC has been an effective resource. The agency should appropriately allocate resources based on critical need and maximum impact to the citrus industry. These citrus health activities directly protect citrus production on approximately 765,000 acres in the United States worth more than $11,000,000,000 in total.

Huanglongbing Multi-Agency Coordination [HLB–MAC] Group

The Committee recognizes the significant economic impact of this disease on the citrus industry, which is especially acute in Florida and a growing concern in both Texas and California. The Committee also understands that growers are requesting the right to try treatments that have begun to show success in early stages of testing. The Committee encourages the HLB–MAC group to explore and identify new methods to expedite the delivery of promising treatments directly to growers. Finally, the Committee expects any funds which are redirected from existing HLB–MAC projects be repurposed to other priority HLB–MAC projects that are showing promising results to ensure these critical funds remain committed to help facilitate the design and implementation of the rapid delivery pathway to growers.

Invasive Tree Pests

The Committee recognizes that the forests products industry and family forest owners are under increasing threat from a growing number of invasive forest pests. It is essential that APHIS carry out a comprehensive program to counter the spread of invasive species and work towards complete eradication of the Asian long-horned beetle. The Secretary is directed to report to the Committee regarding the steps being taken to eradicate the Asian long-horned beetle and spotted lanternfly and to minimize the spread of other pests such as the polyphagous and Kuroshio shot hole borers. As the emerald ash borer continues to spread, APHIS shall continue to assist states that have recent detections of emerald ash borer where assistance will enable states to fully monitor the insect and to inform and manage public and private land owner issues.

National Bio and Agro-Defense Human Capital Development

The Committee notes that significant resources have been invested in the new National Bio and Agro-Defense Facility [NBAF] and is concerned about projected staffing shortages of qualified veterinary diagnosticians and scientists for the NBAF, which is slated for full operation in 2022. The Committee provides $3,000,000 to APHIS to ensure necessary steps are taken to develop a qualified workforce that are subject matter experts in foreign, emerging and zoonotic diseases capable of developing, validating and conducting needed diagnostics, performing epidemiologic studies, and completing bioinformatics analyses.

Non-Lethal Strategies

The Committee is aware that Wildlife Services has deployed the use of non-lethal strategies—including fladry, electric fencing, livestock guardian dogs, alarm systems and lighting—which have effectively reduced predator depredation on livestock in many cases. The Committee encourages Wildlife Services when applicable to (1) proactively work with agricultural producers to implement and evaluate the effectiveness of preventive non-lethal control actions, (2) assess patterns of depredation events and (3) provide deterrent training for agricultural producers, land-owners and Wildlife Services and other agency personnel in non-lethal management strategies in collaboration with the National Wildlife Research Center.

Pacific Ants Prevention Plan

The Secretary is encouraged to lead the revision of the Pacific Ants Prevention Plan, in collaboration with U.S. and international partners. The plan should include: research; the development of technologies and methodologies for prevention, eradication, and control of invasive ants; and the collaborative implementation of projects to prevent, monitor, and control invasive ants in affected Pacific Islands.

[In thousands of dollars]			
	Fiscal year 2018 enacted	**Fiscal year 2019 budget request**	**Committee recommendation**
Appropriations Temporarily Reduced—Sequestration	-77,418	-74,400	-74,400
Budget Authority, Appropriations Act	1,266,582	962,600	1,299,600
Less Obligations:			
Child Nutrition Programs (Entitlement Commodities)	465,000	465,000	485,000
State Option Contract	5,000	5,000	5,000
Removal of Defective Commodities	2,500	2,500	2,500
Disaster Relief	36,000	5,000	5,000
Additional Fruits, Vegetables, and Nuts Purchases	206,000	206,000	206,000
Fresh Fruit and Vegetable Program	171,000	174,000	174,000
Estimated Future Needs	354,524	48,758	48,758
Total, Commodity Procurement	1,210,024	906,258	926,258
Administrative Funds:			
Commodity Purchase Support	35,853	35,853	35,853
Marketing Agreements and Orders	20,705	20,489	20,489
Total, Administrative Funds	56,558	56,342	56,342
Total Obligations	1,266,582	962,600	982,600

Committee Recommendations

The Committee recommends a transfer from section 32 funds of $20,489,000 for the formulation and administration of marketing agreements and orders.

Section 32 Authorities

Under the authority described in clause 3 of 7 U.S.C. 612c, the Secretary is able to direct funds from the section 32 account to increase the purchasing power of producers. This practice has been used on various occasions to provide direct assistance to producers when market forces or

Committee recommendation includes no less than the fiscal year 2018 funding level to improve understanding of the European Strain 1 and North American Strain 1 of the sudden oak death pathogen and treatment methods to inform control and management techniques in wildlands.

West Nile Virus

The Committee remains concerned with the threats to human and animal health posed by West Nile virus and recognizes that a critical strategy for addressing these threats is necessary to prevent the infection and transmission by known vectors, including farm-raised alligators. The Committee encourages APHIS to further investigate West Nile virus and other infectious diseases affecting farm raised alligators and develop treatments and methods to prevent infection and transmission.

Wildlife Damage Management

APHIS is responsible for providing Federal leadership in managing problems caused by wildlife. The Committee provides $108,376,000 for wildlife damage control to maintain priority initiatives, including preventing the transport of invasive snakes and other harmful species. No less than $250,000 should be available for the agency to reduce blackbird depredation in the Northern Great Plains.

The Committee maintains support for assistance to catfish producers to help mitigate wildlife depredation, particularly as it pertains to fish-eating and disease-carrying birds. The Committee provides no less than the fiscal year 2018 level for damage management efforts and the development of methods to assist producers in combatting the persistent threat and economic hardship caused by cormorants, pelicans, and other birds.

The Committee recognizes the importance of the National Feral Swine Damage Management Program in reducing adverse ecological and economic impacts caused by feral swine. The Committee provides no less than the fiscal year 2018 level in support of APHIS efforts to decrease these invasive pests' damage and risk to agriculture, natural resources, and property.

The Committee provides $28,000,000 for the National Rabies Management Program to fortify existing barriers and advance prevention and eradication efforts.

Given the shared and complementary goals of Wildlife Damage Management and the U.S. Fish and Wildlife Services to sustainably integrate wildlife into natural habitats while protecting livestock, the Secretary is directed to coordinate with the U.S. Fish and Wildlife Services on innovative strategies to provide predator management and reduce predator-livestock conflict. The Secretary is further directed to report to this Committee on how the two agencies can work together to improve wildlife management.

Wildlife Services Education and Training

The Committee is aware of the wide range of hazardous procedures and materials utilized by APHIS personnel in the conduct of daily duties. In addition, a recent comprehensive study noted the critical need to provide standardized safety training, certification, and database management for tracking, to ensure the safest working environment possible. As such, the Committee provides $2,000,000 within Wild-life Damage Management to maintain a National Training Academy focused on those areas of greatest concern such as pyrotechnics, firearms, hazardous materials, immobilization and euthanasia drugs, pesticides, animal care and handling, land vehicles, watercraft, and zoonotic diseases.

Wildlife Services Methods Development

The Committee appreciates the important work done by the National Wildlife Research Center and its affiliated field locations to resolve problems caused by the interaction of wild animals and society. The Committee provides $18,856,000 to ensure continued development of technical and scientific information on wildlife damage management.

Buildings and Facilities

Appropriations, 2018	$3,175,000
Budget estimate, 2019	2,852,000
Committee recommendation	3,175,000

The APHIS appropriation for "Buildings and Facilities" funds major nonrecurring construction projects in support of specific program activities and recurring construction, alterations, preventive maintenance, and repairs of existing APHIS facilities.

Committee Recommendations

The Committee recommendation includes an appropriation of $3,175,000 for buildings and facilities of the Animal and Plant Health Inspection Service. This funding is necessary to allow APHIS to maintain existing facilities, and perform critically needed repairs to and replacements of building components, such as heating, ventilation and air-conditioning on a prioritized basis at APHIS facilities. The Committee notes that due to the environmentally sensitive nature of many APHIS facilities, closure of a facility could result if APHIS is unable to complete the required repairs.

Agricultural Marketing Service

Marketing Services

Appropriations, 2018	$151,595,000
Budget estimate, 2019	118,617,000
Committee recommendation	155,845,000

The Agricultural Marketing Service [AMS] was established by the Secretary of Agriculture on April 2, 1972. AMS carries out programs authorized by more than 50 different statutory authorities, the primary ones being the Agricultural Marketing Act of 1946 (7 U.S.C. 1621–1627, 1635–

1638); the U.S. Cotton Standards Act (7 U.S.C. 51–65); the Cotton Statistics and Estimates Act (7 U.S.C. 471–476); the Tobacco Inspection Act (7 U.S.C. 511–511q); the Perishable Agricultural Commodities Act (7 U.S.C. 499a–499t); the Egg Products Inspection Act (21 U.S.C. 1031–1056); and section 32 of the Act of 1935 (Public Law 74–320, 7 U.S.C. 612c).

Programs administered by this agency include the market news services, standardization, grading, classing, shell egg surveillance services, transportation services, wholesale farmers and alternative market development, grant payments to States for marketing activities, the Federal administration of marketing agreements and orders, commodity purchases, Perishable Agricultural Commodities Act, the Plant Variety Protection Act (7 U.S.C. 2321 et seq.), and market protection and promotion activities.

Committee Recommendations

The Committee recommends an appropriation of $155,845,000 for Marketing Services of the Agricultural Marketing Service.

National Organic Program

The Committee provides $15,094,000 for the National Organic Program [NOP], an increase of $3,000,000. A healthy market for organic products requires a clear product distinction backed by a trusted, verified, and enforced label. The Committee recognizes that the NOP, which enforces the organic regulations and ensures they evolve to keep pace with consumer expectations, is essential. In light of recent reports of inadequate enforcement of organic standards, the Committee directs USDA to provide all resources needed for the NOP to deliver the strongest possible oversight before allowing the USDA organic seal to be granted to domestic and international operations and products.

Organic Dairy

The Committee is disappointed by continued reports of inconsistencies in the enforcement and interpretation of regulations that apply to organic dairy farms. The Committee directs the NOP to resolve these issues, and eliminate any inconsistencies in applying and enforcing regulations relating to the transition of livestock to organic dairy production, and also dry matter intake during the grazing season for organic dairy cattle. The Secretary must ensure that organic inspectors, certification file review staff, and NOP Organic Certification staff have documented training and experience in livestock nutrition and grazing on organic dairies with more than 1,000 milking cows if they are certifying operations of that size, and also that separate dry matter intake calculations are made for each category of dairy cow and not averaged among milking and dry cows, and that inspections are conducted during the grazing season.

Organic Data Initiative

The Committee includes $250,000 for AMS to coordinate with NASS for activities related to expanding organic price reporting and organic data collection.

Limitation on Administrative Expenses

Limitation, 2018	$61,227,000
Budget limitation, 2019	60,982,000
Committee recommendation	60,982,000

The Omnibus Budget Reconciliation Act of 1981 (Public Law 97–35) initiated a system of user fees for the cost of grading and classing cotton, and tobacco. These activities, authorized under the U.S. Cotton Standards Act (7 U.S.C. 51 et seq.), the Tobacco Inspection Act (7 U.S.C. 511 et seq.), and other provisions of law are designed to facilitate commerce and protect participants in the industry.

Committee Recommendations

The Committee recommends a limitation of $60,982,000 on administrative expenses of the Agricultural Marketing Service.

Funds for Strengthening Markets, Income, and Supply (Section 32)

(Including Transfers of Funds)	
Appropriations, 2018	$20,705,000
Budget estimate, 2019	20,489,000
Committee recommendation	20,489,000

Under section 32 of the Act of August 24, 1935, (7 U.S.C. 612c), an amount equal to 30 percent of customs receipts collected during each preceding calendar year and unused balances are available for encouraging the domestic consumption and exportation of agricultural commodities. An amount equal to 30 percent of receipts collected on fishery products is transferred to the Department of Commerce. Additional transfers to the child nutrition programs of the Food and Nutrition Service have been provided in recent appropriations Acts. The following table reflects the status of this fund for fiscal years 2018–2019:

Estimated Total Funds Available and Balance Carried Forward — Fiscal Years 2018–2019

[In thousands of dollars]	Fiscal year 2018 enacted	Fiscal year 2019 budget request	Committee recommendation
Appropriation (30% of Customs Receipts)	10,370,878	10,624,198	10,624,198
Less Transfers:			
Food and Nutrition Service	-8,872,010	-9,092,217	-9,092,217
Commerce Department	-154,868	-157,980	-157,980
Total, Transfers	-9,026,878	-9,250,197	-9,250,197
Prior Year Appropriation Available, Start of Year	125,000	125,000	125,000
Transfer of Prior Year Funds to FNS (F&V)	-125,000	-125,000	-125,000
Budget Authority, Farm Bill	1,344,000	1,374,000	1,374,000
Rescission of Current Year Funds		-337,000	

natural conditions adversely affect the financial condition of farmers and ranchers. The Committee notes the importance of the ability of the Secretary to utilize this authority, but believes that communication between the Department and the Congress should be improved when this practice is used. Therefore, the Committee directs the Secretary to provide notification to the Appropriations Committee in advance of any public announcement or release of section 32 funds under the specific authorities cited above.

Payments to States and Possessions

Appropriations, 2018	$1,235,000
Budget estimate, 2019	1,109,000
Committee recommendation	1,235,000

The Federal-State Marketing Improvement Program [FSMIP] is authorized by section 204(b) of the Agricultural Marketing Act of 1946 and is also funded from appropriations. Matching grants are awarded on a competitive basis to State marketing agencies to identify and test market alternative farm commodities, determine methods of providing more reliable market information, and develop better commodity grading standards. This program has made possible many types of projects, such as electronic marketing and agricultural product diversification. Current projects are focused on the improvement of marketing efficiency and effectiveness, and seeking new outlets for existing farm produced commodities. The legislation grants the U.S. Department of Agriculture authority to establish cooperative agreements with State departments of agriculture or similar State agencies to improve the efficiency of the agricultural marketing chain. The States perform the work or contract it to others, and must contribute at least one-half of the cost of the projects.

Committee Recommendations

The Committee recommends an appropriation of $1,235,000 for Payments to States and Possessions for Federal-State marketing projects and activities.

Limitation on Inspection and Weighing Services Expenses

Limitation, 2018	$55,000,000
Budget limitation, 2019	80,000,000
Committee recommendation	55,000,000

The agency provides an official grain inspection and weighing system under the U.S. Grain Standards Act [USGSA], and official inspection of rice and grain-related products under the Agricultural Marketing Act [AMA] of 1946. The USGSA was amended in 1981 to require the collection of user fees to fund the costs associated with the operation, supervision, and administration of Federal grain inspection and weighing activities.

Committee Recommendations

The Committee recommends a limitation of $55,000,000 on inspection and weighing services expenses.

Office of the Under Secretary for Food Safety

Appropriations, 2018	$800,000
Budget estimate, 2019	800,000
Committee recommendation	800,000

The Office of the Under Secretary for Food Safety provides direction and coordination in carrying out the laws enacted by the Congress with respect to the Department's inspection of meat, poultry, and processed egg products. The Office has oversight and management responsibilities for the Food Safety and Inspection Service.

Committee Recommendations

The Committee recommends an appropriation of $800,000 for the Office of the Under Secretary for Food Safety.

Food Safety and Inspection Service

Appropriations, 2018	$1,056,844,000
Budget estimate, 2019	1,032,273,000
Committee recommendation	1,049,344,000

The major objectives of the Food Safety and Inspection Service are to assure that meat and poultry products are wholesome, unadulterated, and properly labeled and packaged, as required by the Federal Meat Inspection Act (21 U.S.C. 601 et seq.) and the Poultry Products Inspection Act (21 U.S.C. 451 et seq.); and to provide continuous in-plant inspection to egg processing plants under the Egg Products Inspection Act.

The Food Safety and Inspection Service was established on June 17, 1981, by Secretary's Memorandum No. 1000–1, issued pursuant to Reorganization Plan No. 2 of 1953.

The inspection program of the Food Safety and Inspection Service provides in-plant inspection of all domestic plants preparing meat, poultry or egg products for sale or distribution; reviews foreign inspection systems and establishments that prepare meat or poultry products for export to the United States; and provides technical and financial assistance to States which maintain meat and poultry inspection programs.

Committee Recommendations

The Committee recommends an appropriation of $1,049,344,000 for the Food Safety and Inspection Service [FSIS].

Humane Slaughter

The Committee directs FSIS to continue to provide annual reports to the Committee on the implementation of objective scoring methods undertaken by FSIS to enforce the Humane Methods of Slaughter Act.

The Committee also directs FSIS to ensure that personnel hired with funding previously provided specifically for Humane Methods of Slaughter Act enforcement focus their attention on overseeing compliance with humane handling rules for live animals as they arrive and are offloaded and handled in pens, chutes, and stunning areas and that all inspectors receive robust training.

The following table represents the Committee's specific recommendations for the Food Safety and Inspection Service as compared to the fiscal year 2018 and budget request levels:

Food Safety and Inspection Service Salaries and Expenses

[In thousands of dollars]			
	Fiscal year 2018 enacted	**Fiscal year 2019 budget request**	**Committee recommendation**
Food safety inspection:			
Federal	943,824	920,403	936,324
State	61,682	61,150	61,682
International	16,758	16,375	16,758
PHDCIS	34,580	34,345	34,580
Total	1,056,844	1,032,273	1,049,344

TITLE II: FARM PRODUCTION AND CONSERVATION PROGRAMS

Office of the Under Secretary for Farm Production and Conservation

Appropriations, 2018	$901,000
Budget estimate, 2019	875,000
Committee recommendation	901,000

The Office of the Under Secretary for Farm Production and Conservation provides direction and coordination in carrying out the laws enacted by the Congress with respect to the Department's commodity programs, farm loans, disaster assistance, crop insurance, natural resources conservation and environment programs, and some energy programs. The Office has oversight and management responsibilities for the Farm Service Agency (including the Commodity Credit Corporation), Risk Management Agency, and the Natural Resources Conservation Service.

Committee Recommendations

The Committee recommends an appropriation of $901,000 for the Office of the Under Secretary for Farm Production and Conservation.

Emergency Response

The Committee directs USDA to produce a report outlining the average and longest length of time it takes USDA to provide reimbursement under the following emergency assistance programs: crop insurance; Noninsured Crop Disaster Assistance Program [NAP]; Livestock Indemnity Program [LIP]; Live-stock Forage Disaster Program [LFP]; Emergency Assistance for Livestock, Honey Bees, and Farm-Raised Fish Program [ELAP]; Tree Assistance Program [TAP]; Emergency Conservation Program; and Emergency Forest Restoration Program [EFRP]. USDA is also directed to include in the report any barriers to implementing a more efficient reimbursement process and recommendations to the Committee on potential improvements.

Farm Production and Conservation Business Center

Appropriations, 2018	$1,028,000
Budget estimate, 2019	196,402,000
Committee recommendation	1,028,000

The Committee recommends an appropriation of $1,028,000 for the Farm Production and Conservation Business Center.

The Committee notes that USDA has yet to finalize and submit a design plan for the Farm Production and Conservation (FPAC) Business Center, and the true cost estimates and staffing requirements for implementation for fiscal year 2019 are still unknown. While the Committee supports the streamlined efficiencies of the FPAC mission area and provided funding accordingly in fiscal year 2018, the proposed budget increase for the Business Center is premature until a final design plan is completed. The Committee further notes that the Secretary maintains interchange authority in the interim.

Farm Service Agency

The Farm Service Agency [FSA] was established October 13, 1994, pursuant to the Federal Crop Insurance Reform and Department of Agriculture Reorganization Act of 1994, Public Law 103–354. The FSA administers a variety of activities, such as the commodity price support and production adjustment programs financed by the Commodity Credit Corporation; the Conservation Reserve Program [CRP]; the Emergency Conservation Program; the Commodity Operation Programs including the warehouse examination function; farm ownership, farm operating, emergency disaster, and other loan programs; and the Noninsured Crop Disaster Assistance Program [NAP], which provides crop loss protection for growers of many crops for which crop insurance is not available. In addition, FSA currently provides certain administrative support services to the Foreign Agricultural Service [FAS] and to the Risk Management Agency [RMA].

Salaries and Expenses

The account Salaries and Expenses, Farm Service Agency, funds the administrative expenses of program administration and other functions assigned to FSA. The funds consist of appropriations and transfers from the CCC export credit guarantees, Food for Peace loans, and Agricultural Credit Insurance Fund program accounts, and miscellaneous advances from other sources. All administrative funds used by FSA are consolidated into one account. The consolidation provides clarity and better management and control of funds, and facilitates accounting, fiscal, and budgetary work by eliminating the necessity for making individual allocations and allotments and maintaining and recording obligations and expenditures under numerous separate accounts.

(Including Transfers of Funds) [In thousands of dollars]			
	Appropriations	**Transfers from program accounts**	**Total, FSA, salaries and expenses**
Appropriations, 2018	1,202,146	317,610	1,519,756
Budget estimate, 2019	920,490	266,913	1,187,403
Committee recommendation	1,202,146	317,603	1,519,749

Committee Recommendations

The Committee recommends an appropriation of $1,519,749,000 for salaries and expenses of the Farm Service Agency, including a direct appropriation of $1,202,146,000. The Committee supports the mission of FSA and the important services that they provide across the country; therefore, the Committee does not accept the full decrease for information technology as proposed in the budget. The Committee provides $5,000,000 for the Farm Production and Conservation Business Center information portal.

Continuous Conservation Reserve Program

The Secretary is strongly encouraged to, within the total acreage made available for enrollment in the conservation reserve program and without reducing the periodic availability of general signup, enroll, to the maximum extent practicable, acreage for activities included in the State Acres for Wildlife Enhancement practice or other similar administratively established wetland and habitat practices that benefit priority fish and wildlife species identified in State, regional, and national conservation initiatives with a priority for initiatives that provide large blocks of cover ideal for wildlife nesting.

Risk Management Agency

Salaries and Expenses

Appropriations, 2018	$74,829,000
Budget estimate, 2019	37,942,000
Committee recommendation	74,829,000

The Risk Management Agency performs administrative functions relative to the Federal crop insurance program that is authorized by the Federal Crop Insurance Act (7 U.S.C. 1508), as amended by the Agricultural Risk Protection Act of 2000 [ARPA], Public Law 106–224, and the Agricultural Act of 2014 (Public Law 113–79).

ARPA authorized significant changes in the crop insurance program. This act provides higher government subsidies for producer premiums to make coverage more affordable; expands research and development for new insurance products and under-served areas through contracts with the private sector; and tightens compliance. Functional areas of risk management are: research and development; insurance services; and compliance, whose functions include policy formulation and procedures and regulations development.

Committee Recommendations

The Committee recommends an appropriation of $74,829,000 for Risk Management Agency, Salaries and Expenses.

The Committee recognizes that there are many research priorities that competitive funding may be used to address, including the feasibility of insurance programs to cover business interruption due to integrator bankruptcy and catastrophic loss in the poultry industry. The Committee encourages RMA to support research into these priorities.

Information Technology

The Committee remains dedicated to ensuring FSA has reliable and functioning IT systems because it is critical that farmers and ranchers have access to the tools they need to succeed. The Committee has invested significant taxpayer dollars to modernize outdated systems and continues to provide resources above the budget request. The Committee continues statutory language that allows funds for IT to be obligated only after the Secretary meets certain reporting requirements. The Committee has reviewed the third-party IT analysis and expects the agency to follow the recommendations where applicable.

National Agriculture Imagery Program

The Committee recommends that funding shall be allocated to purchase imagery products to meet programmatic requirements.

Non-Insured Crop Disaster Assistance

The Committee is concerned that the Department is failing to provide adequate risk protection through the Non-insured Crop Assistance Program for farmers producing hay forage with legume grass mixes. These crops often suffer quality loses resulting in adverse economic impacts, yet these farms have no meaningful source of risk protection. The Committee encourages the Secretary to establish within the Non-insured Crop Assistance Program an additional Forage Category to reflect the relative feed value, and the potential for quality loss, of legume grass mix forage.

State Mediation Grants

Appropriations, 2018	$3,904,000
Budget estimate, 2019	3,228,000
Committee recommendation	3,904,000

This program is authorized under title V of the Agricultural Credit Act of 1987 (7 U.S.C. 5101 et seq.). Originally designed to address agricultural credit disputes, the program was expanded by the Federal Crop Insurance

Reform and Department of Agriculture Reorganization Act of 1994 (Public Law 103–354) to include other agricultural issues such as wetland determinations, conservation compliance, rural water loan programs, grazing on National Forest System lands, and pesticides. Grants are made to States whose mediation programs have been certified by the FSA. Grants will be solely for operation and administration of the State's agricultural mediation program.

Committee Recommendations

The Committee recommends an appropriation of $3,904,000 for State Mediation Grants.

Grassroots Source Water Protection Program

Appropriations, 2018	$6,500,000
Budget estimate, 2019	
Committee recommendation	6,500,000

This program is intended to assist in the protection of ground-water through State rural water associations.

Committee Recommendations

The Committee recommends an appropriation of $6,500,000 for Grassroots Source Water Protection.

Dairy Indemnity Program

(Including Transfers of Funds)	
Appropriations, 2018	$500,000
Budget estimate, 2019	500,000
Committee recommendation	500,000

Under the program, the Department makes indemnification payments to dairy farmers and manufacturers of dairy products who, through no fault of their own, suffer income losses because they are directed to remove their milk from commercial markets due to contamination of their products by registered pesticides. The program also authorizes indemnity payments to dairy farmers for losses resulting from the removal of cows or dairy products from the market due to nuclear radiation or fallout.

Committee Recommendations

The Committee recommends an appropriation of such sums as may be necessary, estimated in fiscal year 2019 to be $500,000, for indemnity payments to dairy farmers.

Agricultural Credit Insurance Fund Program Account (Including Transfers of Funds)

The Agricultural Credit Insurance Fund Program Account is used to provide direct and guaranteed farm ownership, farm operating, conservation, Indian highly fractioned land, and emergency loans to individuals, as well as the following types of loans to associations: irrigation and drainage, grazing, Indian tribe land acquisition, and boll weevil eradication.

FSA is also authorized to provide financial assistance to borrowers by guaranteeing loans made by private lenders having a contract of guarantee from FSA as approved by the Secretary of Agriculture and to establish Beginning Farmer and Rancher Individual Development grant accounts.

The following programs are financed through this fund:

Boll Weevil Eradication Loans

Made to assist foundations in financing the operations of the boll weevil eradication programs provided to farmers.

Credit Sales of Acquired Property

Property is sold out of inventory and is made available to an eligible buyer by providing FSA loans.

Emergency Loans

Made to producers to aid recovery from production and physical losses due to drought, flooding, other natural disasters, or quarantine. The loans may be used to: restore or replace essential property; pay all or part of production costs associated with the disaster year; pay essential family living expenses; reorganize the farming operation; and refinance certain debts.

Farm Operating Loans

Provide short-to-intermediate term production or chattel credit to farmers who cannot obtain credit elsewhere, to improve their farm and home operations, and to develop or maintain a reasonable standard of living. The term of the loan varies from 1 to 7 years.

Farm Ownership Loans

Made to borrowers who cannot obtain credit elsewhere to restructure their debts, improve or purchase farms, refinance nonfarm enterprises which supplement but do not supplant farm income, or make additions to farms. Loans are made for 40 years or less.

Indian Tribe Land Acquisition Loans

Made to any Indian tribe recognized by the Secretary of the Interior or tribal corporation established pursuant to the Indian Reorganization Act

(Public Law 93–638) which does not have adequate uncommitted funds to acquire lands or interest in lands within the tribe's reservation or Alaskan Indian community, as determined by the Secretary of the Interior, for use of the tribe or the corporation or the members thereof.

Indian Highly Fractionated Land Loans

Made to Indian tribal members to purchase highly fractionated lands, as authorized by the Food, Conservation, and Energy Act of 2008.

Committee Recommendations

The Committee recommends a total loan level of $8,017,668,000 for programs within the Agricultural Credit Insurance Fund Program Account.

Loan Programs

The Committee continues to support FSA loan programs that ensure farmers and ranchers have access to credit to maintain and improve their operations. The Committee is aware of the heightened operating loan activity in fiscal year 2018 and notes the statutory authority allowing program level increases that do not require additional budget authority. The Committee will continue to monitor program demand in the coming months and directs FSA to provide timely estimates for future needs.

The following table reflects the program levels for farm credit programs administered by the Farm Service Agency recommended by the Committee, as compared to the fiscal year 2018 and the budget request levels:

Agricultural Credit Programs

Loan Levels

The Federal Credit Reform Act of 1990 established the program account. Appropriations to this account are used to cover the life-time subsidy costs associated with the direct loans obligated and loan guarantees committed, as well as for administrative expenses.

[In thousands of dollars]			
	Fiscal year 2018 enacted	Fiscal year 2019 budget request	Committee recommendation
Farm Ownership:			
Direct	1,500,000	1,500,000	1,500,000
Guaranteed	2,750,000	2,750,000	2,750,000
Farm Operating:			
Direct	1,530,000	1,500,000	1,530,000
Guaranteed unsubsidized	1,960,000	1,600,000	1,960,000
Emergency Loans	25,610	37,668	37,668
Indian Tribe Land Acquisition	20,000	20,000	20,000
Conservation Loans:			
Guaranteed	150,000	150,000	150,000
Indian Highly Fractionated Land Loans	10,000		10,000
Boll Weevil Eradication	60,000	60,000	60,000
Total, Loan Authorizations	8,005,610	7,617,668	8,017,668

The following table reflects the cost of programs under credit reform:

[In thousands of dollars]			
	Fiscal year 2018 enacted	Fiscal year 2019 budget request	Committee recommendation
Farm Operating:			
Direct	61,812	58,500	59,670
Guaranteed unsubsidized	21,756	17,280	21,168
Emergency Loans	1,260	1,567	1,567
Indian Highly Fractionated Land Loans	2,272		2,134
Total, Loan Subsidies	87,100	77,347	84,539
ACIF Expenses:			
Salaries and Expenses	314,998	226,436	314,998
Administrative Expenses	10,070	10,070	10,070
Total, ACIF Expenses	325,068	292,587	325,068

Natural Resources Conservation Service

The Natural Resources Conservation Service [NRCS] was established pursuant to Public Law 103–354, the Department of Agriculture Reorganization Act of 1994 (7 U.S.C. 6962). The NRCS works with conservation districts, watershed groups, and Federal and State agencies to bring about physical adjustments in land use that will conserve soil and water resources, provide for agricultural production on a sustained basis, and reduce flood damage and sedimentation.

Conservation Operations

Appropriations, 2018	$874,107,000
Budget estimate, 2019	669,033,000
Committee recommendation	879,107,000

Conservation operations are authorized by Public Law 74–46 (16 U.S.C. 590a–590f). Activities include:

Conservation Technical Assistance provides assistance to district cooperators and other land users in the planning and application of conservation treatments to control erosion and improve the quantity and quality of soil resources, improve and conserve water, enhance fish and wildlife habitat, conserve energy, improve wood-land, pasture and range conditions, and reduce upstream flooding; all to protect and enhance the natural resource base.

Resource appraisal and program development ensures that programs administered by the Secretary of Agriculture for the conservation of soil, water, and related resources shall respond to the Nation's long-term needs.

Plant Materials Centers assemble, test, and encourage increased use of plant species which show promise for use in the treatment of conservation problem areas.

Snow Survey and Water Forecasting provides estimates of annual water availability from high mountain snow packs and relates to summer stream flow in the Western States and Alaska. Information is used by agriculture, industry, and cities in estimating future water supplies.

Soil Surveys inventory the Nation's basic soil resources and determine land capabilities and conservation treatment needs. Soil survey publications include interpretations useful to cooperators, other Federal agencies, State, and local organizations.

Committee Recommendations

The Committee recommends an appropriation of $879,107,000 for Conservation Operations. The Committee recommendation includes $773,844,000 for Conservation Technical Assistance, $80,802,000 for Soil Surveys, $9,380,000 for Snow Survey and Water Forecasting, and $9,481,000 for Plant Materials Centers.

Acre-for-Acre Wetlands Mitigation

The Secretary is encouraged to use mitigation with the conversion of a natural wetland and equivalent wetlands functions at a ratio not to exceed a ratio of 1-to-1 acreage.

Agricultural Management Assistance

In carrying out the programs under section 524(b) of the Federal Crop Insurance Act, the Secretary is encouraged to establish multi-year pilot projects to provide financial and technical assistance to farms regulated under the FSMA Produce Safety Rule, for capital improvements required to comply with the rule and otherwise enhance food safety practices at the production level. Payment limits and other provisions of the AMA program will apply.

Conservation Practices

The Committee encourages USDA's Natural Resources Conservation Service to prioritize funding Environmental Quality Incentives Program practices that score highly on the Conservation Practices Physical Effects [CPPE] matrix.

NRCS Staffing

NRCS is directed to provide an update on staffing levels at NRCS offices across the country, including vacancies that have remained unfilled for more than 6 months, and plans to fill those vacancies. The Committee is concerned that unfilled state-level positions are creating delays in application approval and the deployment of important conservation funding.

Program Duplication

The Committee directs NRCS to provide a report within 90 days of enactment of this Act on actions it will take to eliminate program duplication as identified in IG reports.

Technical Assistance

The Committee directs NRCS to maintain a record of total technical assistance dollars for the past 3 years and annually in the future, and provide the data to the Appropriations Subcommittee on Agriculture, and the Committee on Agriculture, Nutrition and Forestry. This chapter should differentiate mandatory and discretionary allocations.

Watershed and Flood Prevention Operations

Appropriations, 2018	$150,000,000
Budget estimate, 2019	
Committee recommendation	150,000,000

The Watershed Protection and Flood Prevention Act (Public Law 566, 83d Cong.) (16 U.S.C. 1000–1005, 1007–1009) provides for cooperation between the Federal Government and the States and their political subdivisions in a program to prevent erosion, flood-water, and sediment damages in the watersheds or rivers and streams and to further the conservation, development, utilization, and disposal of water, and to further the conservation and proper utilization of land in authorized watersheds.

Committee Recommendations

The Committee recommends an appropriation of $150,000,000 for the Watershed and Flood Prevention Operations Program.

The Committee recognizes the critical challenges facing rural water resource management and protection and supports needed investments in watershed operations. These Federal-State-local partnerships are uniquely positioned to identify critical watershed protection and flood prevention needs in rural communities and implement projects that deliver multiple streams of benefits for homes, businesses, transportation infrastructure, and natural resources. In selecting projects for funding, the Committee expects the agency to balance the needs of addressing the project backlog, remediation of existing structures, and new projects.

Watershed Rehabilitation Program

Appropriations, 2018	$10,000,000
Budget estimate, 2019	
Committee recommendation	

The Watershed Rehabilitation Program account provides for technical and financial assistance to carry out rehabilitation of structural measures, in accordance with section 14 of the Watershed Protection and Flood Prevention Act, approved August 4, 1954 (16 U.S.C. 1012, U.S.C. 1001, et seq.), as amended by section 313 of Public Law 106–472, November 9, 2000, and by section 2803 of Public Law 110–246.

Committee Recommendations

The Committee does not recommend an appropriation for the Watershed Rehabilitation Program.

Corporations: Federal Crop Insurance Corporation Fund

Appropriations, 2018	$8,913,000,000
Budget estimate, 2019	8,687,000,000
Committee recommendation	8,687,000,000

The Federal Crop Insurance Act, as amended by the Federal Crop Insurance Reform Act of 1994, authorizes the payment of expenses which may include indemnity payments, loss adjustment, delivery expenses, program-related research and development, startup costs for implementing this legislation such as studies, pilot projects, data processing improvements, public outreach, and related tasks and functions.

All program costs, except for Federal salaries and expenses, are mandatory expenditures subject to appropriation.

Committee Recommendations

The Committee recommends an appropriation of such sums as may be necessary, estimated to be $8,687,000,000 in fiscal year 2019 for the Federal Crop Insurance Corporation Fund.

Commodity Credit Corporation Fund

The Commodity Credit Corporation [CCC] is a wholly owned Government corporation created in 1933 to stabilize, support, and protect farm income and prices; to help maintain balanced and adequate supplies of agricultural commodities, including products, foods, feeds, and fibers; and to help in the orderly distribution of these commodities. CCC was originally incorporated under a Delaware charter and was reincorporated June 30, 1948, as a Federal corporation within the Department of Agriculture by the Commodity Credit Corporation Charter Act, approved June 29, 1948 (15 U.S.C. 714).

The Commodity Credit Corporation engages in buying, selling, lending, and other activities with respect to agricultural commodities, their products, food, feed, and fibers. Its purposes include stabilizing, supporting, and protecting farm income and prices; maintaining the balance and adequate supplies of selected commodities; and facilitating the orderly distribution of such commodities. In addition, the Corporation makes available materials and facilities required in connection with the storage and distribution of such commodities. The Corporation also disburses funds for sharing of costs with producers for the establishment of approved conservation practices on environmentally sensitive land and subsequent rental payments for such land for the duration of Conservation Reserve Program contracts.

Corporation activities are primarily governed by the following statutes: the Commodity Credit Corporation Charter Act (Public Law 80–806), as amended; the Agricultural Act of 1949 (Public Law 81–439), as amended (1949 Act); the Agricultural Adjustment Act of 1938 (Public Law 75–430), as amended (the 1938 Act); the Food Security Act of 1985 (Public Law 99–198), as amended (1985 Act); the Food, Conservation, and Energy Act of 2008 (Public Law 110–246); and the Agricultural Act of 2014 (Public Law 113–79).

Management of the Corporation is vested in a board of directors, subject to the general supervision and direction of the Secretary of Agriculture, who is an ex officio director and chairman of the board. The board consists of seven members, in addition to the Secretary, who are appointed by the President of the United States with the advice and consent of the Senate. Officers of the Corporation are designated according to their positions in the Department of Agriculture.

The activities of the Corporation are carried out mainly by the personnel and through the facilities of the Farm Service Agency [FSA] and the Farm Service Agency State and county committees. The Foreign Agricultural Service, the General Sales Manager, other agencies and offices of the Department, and commercial agents are also used to carry out certain aspects of the Corporation's activities.

Under Public Law 87–155 (15 U.S.C. 713a–11, 713a–12), annual appropriations are authorized for each fiscal year, commencing with fiscal year 1961. These appropriations are to reimburse the Corporation for net realized losses.

Reimbursement for Net Realized Losses

(Including Transfers of Funds)	
Appropriations, 2018	$14,284,847,000
Budget estimate, 2019	15,410,000,000
Committee recommendation	15,410,000,000

Committee Recommendations

The Committee recommends an appropriation of such sums as may be necessary, estimated in fiscal year 2019 to be $15,410,000,000, for the payment to reimburse the Commodity Credit Corporation for net realized losses.

CRP Wetland Restoration and Wildlife Enhancement

The Committee notes that agricultural commodity crops, if left unharvested, may help reduce degradation of wetlands and improve sediment trapping, surface and ground water supply, erosion control, and wildlife habitat while providing winter food for waterfowl and other wildlife. The Committee directs the Commodity Credit Corporation, within 60 days of enactment, to amend its program policies and guidelines for CRP conservation practices CP23 and CP23A, to provide that current and future participants are permitted to plant— but not harvest—agricultural commodity crops as wildlife food plots on up to 10 percent of the enrolled land to enhance waterfowl and upland bird food and habitat.

Hazardous Waste Management

(Limitation on Expenses)	
Limitation, 2018	$5,000,000
Budget estimate, 2019	5,000,000
Committee recommendation	5,000,000

The Commodity Credit Corporation's [CCC] hazardous waste management program is intended to ensure compliance with the Comprehensive Environmental Response, Compensation, and Liability Act (42 U.S.C. 9601 et seq.) and the Resource Conservation and Recovery Act (42 U.S.C. 6901 et seq.). The CCC funds operations and maintenance costs as well as site investigation and cleanup expenses. Investigative and cleanup costs associated with the management of CCC hazardous waste are also paid from USDA's hazardous waste management appropriation.

Committee Recommendations

The Committee recommends a limitation of $5,000,000 for the Commodity Credit Corporation's hazardous waste management program.

TITLE III: RURAL DEVELOPMENT PROGRAMS

The Federal Crop Insurance Reform and Department of Agriculture Reorganization Act of 1994 (Public Law 103–354) abolished the Farmers Home Administration, Rural Development Administration, and Rural Electrification Administration and replaced those agencies with the Rural Housing and Community Development Service, (currently, the Rural Housing Service), Rural Business and Cooperative Development Service (currently, the Rural Business—Cooperative Service), and Rural Utilities Service and placed them under the oversight of the Under Secretary for Rural Economic and Community Development, (currently, Rural Development). These agencies deliver a variety of programs through a network of State and field offices.

Rural Development

Salaries and Expenses

These funds are used to administer the loan and grant programs of the Rural Utilities Service, the Rural Housing Service, and the Rural

Business—Cooperative Service, including reviewing applications, making and collecting loans and providing technical assistance and guidance to borrowers; and to assist in extending other Federal programs to people in rural areas.

(Including Transfers of Funds) [In thousands of dollars]			
	Fiscal year 2018 enacted	**Fiscal year 2019 budget request**	**Committee recommendation**
Appropriation	230,835	156,054	232,835
Transfer from:			
Rural Housing Insurance Fund Loan Program Account	412,254	244,249	412,254
Rural Electrification and Telecommunications Program Account	33,270	38,027	33,270
Rural Development Loan Program Account	4,468		4,468
Community Facilities Program Account		147,591	
Distance Learning, Telemedicine and Broadband Program Account		8,057	
Total, Rural Development salaries and expenses	680,827	612,127	682,827

Under credit reform, administrative costs associated with loan programs are appropriated to the program accounts. Appropriations to the salaries and expenses account will be for costs associated with grant programs.

Committee Recommendations

The Committee recommends $682,827,000 for salaries and expenses of Rural Development.

Bioenergy Program for Advanced Biofuels

The Committee is concerned with the interim rule proposed by the Department under the Bioenergy Program for Advanced Biofuels program (section 9005 of the Energy title of the farm bill, Public Law 113–79), which is intended to promote the development of different qualifying advanced fuel categories. The Committee is concerned that the allocation formula for distribution of section 9005 funds among the qualified fuel categories is inequitable, disproportionate, and inconsistent with the purpose and intent of the section 9005 program. The Committee urges the Department to administer the section 9005 program in a way that is fuel and technology-neutral. Consistent with these objectives, the Committee directs USDA to propose amendments to the interim rule to ensure that any final rule to implement section 9005 provides for a more equitable and proportional allocation of funding among the qualified advanced biofuels and the energy pathways they represent.

Valuing Collateral

The Committee directs RD to provide technical assistance to participating lenders in valuing collateral, including but not limited to telecommunications infrastructure, for the purposes of the Rural Development loan guarantee programs.

Rural Housing Service

The Rural Housing Service [RHS] was established under the Federal Crop Insurance Reform and Department of Agriculture Reorganization Act of 1994, dated October 13, 1994.

Rural Housing Insurance Fund Program Account

(Including Transfers of Funds)	
Appropriations, 2018 (budget authority)	$483,716,000
Budget estimate, 2019 (budget authority)	244,249,000
Committee recommendation (budget authority)	493,945,000

This fund was established in 1965 (Public Law 89–117) pursuant to section 517 of title V of the Housing Act of 1949 (42 U.S.C. 517(d)), as amended. This fund may be used to insure or guarantee rural housing loans for single-family homes, rental and cooperative housing, farm labor housing, and rural housing sites. Rural housing loans are made to construct, improve, alter, repair, or replace dwellings and essential farm service buildings that are modest in size, design, and cost. Rental housing insured loans are made to individuals, corporations, associations, trusts, or partnerships to provide low-cost rental housing and related facilities in rural areas. These loans are repayable in terms up to 30 years.

Committee Recommendations

The Committee recommends an appropriation of $493,945,000 for the Rural Housing Insurance Fund Program Account [RHIF].

The Federal Credit Reform Act of 1990 (Public Law 101–508) established the RHIF program account. Appropriations to this account will be used to cover the lifetime subsidy costs associated with the direct loans obligated and loan guarantees committed in 2019, as well as for administrative expenses. The following table presents the loan subsidy levels as compared to the 2018 levels and the 2019 budget request:

Rural Housing Insurance Fund Program Account

[In thousands of dollars]			
	Fiscal year 2018 enacted	**Fiscal year 2019 budget request**	**Committee recommendation**
Loan Levels:			
Single-Family Housing (sec. 502):			
Direct	1,100,000		1,100,000
Guaranteed	24,000,000	24,000,000	24,000,000
Housing repair (sec. 504)	28,000		28,000
Direct rental housing (sec. 515)	40,000		40,000
Guaranteed rental housing (sec. 538)	230,000	250,000	230,000
Site development loans (sec. 524)	5,000		5,000

[In thousands of dollars]			
	Fiscal year 2018 enacted	**Fiscal year 2019 budget request**	**Committee recommendation**
Credit sales of acquired property	10,000	10,000	10,000
Self help land development loans (sec. 523)	5,000		5,000
Farm labor housing loans (sec. 514)	23,855		23,855
Total, loan levels	25,441,855	24,260,000	25,441,855
Loan Subsidies and Grants:			
Single-Family Housing (sec. 502):			
Direct	42,350		53,900
Housing repair (sec. 504)	3,452		3,419
Direct rental housing (sec. 515)	10,524		9,484
Site development (sec. 524)	58		176
Self help land development (sec. 523)	368		431
Farm labor housing loans (sec. 514)	6,374		5,945
Farm labor housing grants (sec. 516)	8,336		8,336
Total, loan subsidies and grants	71,462		81,691
Administrative expenses	412,254	244,249	412,254
Total, loan subsidies and administrative expenses	483,716	244,249	493,945

Energy Efficiency

The Committee recognizes opportunities to reduce costs for rural housing and save taxpayer money by embracing energy efficiency standards in rural housing, with measures such as air sealing, and installing insulation, window films, and roofs.

Maturing Mortgages

The Committee remains concerned about the alarming number of multi-family housing mortgages scheduled to mature in the next few years. As these mortgages mature, projects and units will be removed from USDA's affordable rural housing program, placing very low income rural residents in jeopardy of untenable rent increases and possible eviction.

The Secretary is directed to engage affordable housing advocates, project owners, tenants, and others as practicable, to find acceptable and effective long term solutions that will retain projects in the affordable rural housing program.

Rural Housing

The Committee is concerned that Rural Housing Service resources do not reach minority, socially disadvantaged, and tribal communities in proportion to their rural populations. For example in fiscal year 2016, of the 7,113 direct loans made nationally by the Rural Housing Service, only 12 went to American Indians or Alaska Natives on tribal land. The Committee believes that the use of Rural Housing loans or funding should support minority, socially disadvantaged, tribal communities, and all other eligible applicants, so as to ensure they have equal access to Rural Housing programs. The Committee directs the Secretary to work with tribal governments and native community development financial institutions (Native CDFIs) to better prepare American Indians or Alaska Natives applicants on tribal land. The Committee further directs USDA Rural Housing Service to report to the Committee on the distribution of loans and resources to these three populations no later than 90 days after enactment of this act.

Rental Assistance Program

Appropriations, 2018	$1,345,293,000
Budget estimate, 2019	1,331,400,000
Committee recommendation	1,331,400,000

Rental assistance is authorized under section 521(a)(2) of the Housing Act of 1949, as amended (42 U.S.C. 1490a). The objective of the program is to reduce rents paid by low-income families living in Rural Housing Service financed rental projects and farm labor housing projects. Under this program, low-income tenants will contribute the higher of: (1) 30 percent of monthly adjusted income; (2) 10 percent of monthly income; or (3) designated housing payments from a welfare agency.

Payments from the fund are made to the project owner for the difference between the tenant's payment and the approved rental rate established for the unit.

The program is administered in tandem with the Rural Housing Service section 515 rural rental housing program and the farm labor loan and grant programs. Priority is given to existing projects for units occupied by rent over-burdened low-income families and projects experiencing financial difficulties beyond the control of the owner.

Committee Recommendations

The Committee recommends an appropriation of $1,331,400,000 for the Rental Assistance Program.

Multi-Family Housing Revitalization Program Account

Appropriations, 2018	$47,000,000
Budget estimate, 2019	
Committee recommendation	50,000,000

The Rural Housing Voucher Program was authorized under the Housing Act of 1949 (42 U.S.C. 1940r) to assist very low income families and individuals who reside in rental housing in rural areas. Housing vouchers may be provided to residents of rental housing projects financed by section 515 loans that have been prepaid after September 30, 2005. Voucher amounts reflect the difference between comparable market rents and tenant-paid rent prior to loan prepayment. Vouchers allow tenants to remain in existing projects or move to other rental housing.

The Multi-family Housing Revitalization Program includes funding for housing vouchers and a demonstration program for the preservation and revitalization of affordable multi-family housing projects. Rural Development's multi-family housing portfolio faces dual pressures for loan prepayments and repair/rehabilitation stemming from inadequate reserves resulting in deferred property maintenance.

Provision of affordable rental housing can be accomplished more economically by revitalizing existing housing stock rather than funding new construction. The Multi-family Housing Revitalization Program includes revitalization tools for maintenance of existing units and vouchers to protect tenants in those projects that prepay. Flexibility is provided to allow Rural Development to utilize funding to meet the most urgent local needs for tenant protection and project revitalization.

Committee Recommendations

The Committee recommends an appropriation of $50,000,000 for the Multi-family Housing Revitalization Program, including $26,000,000 for vouchers and $24,000,000 for a housing preservation demonstration program.

Multi-Family Housing Preservation

The Committee directs the Secretary to provide a report within 120 days of enactment of this Act to estimate the cost of providing rural housing vouchers to all low income households currently receiving USDA rental assistance and residing in a property financed with a Section 515 loan that are set to mature in the subsequent fiscal year and subsequent ten fiscal years. In addition, the Secretary is directed to provide quarterly reports to the Committee on transfers between vouchers and the housing preservation demonstration program within the Multi-Family Housing Revitalization Program Account.

Mutual and Self-Help Housing Grants

Appropriations, 2018	$30,000,000
Budget estimate, 2019	
Committee recommendation	30,000,000

The Mutual and Self-Help Housing Grants Program is authorized by title V of the Housing Act of 1949. Grants are made to local organizations to promote the development of mutual or self-help programs under which groups of usually 6 to 10 families build their own homes by mutually exchanging labor. Funds may be used to pay the cost of construction supervisors who work with families in the construction of their homes and for administrative expenses of the organizations providing the self-help assistance.

Committee Recommendations

The Committee recommends an appropriation of $30,000,000 for Mutual and Self-Help Housing Grants.

Rural Housing Assistance Grants

Appropriations, 2018	$40,000,000
Budget estimate, 2019	
Committee recommendation	40,000,000

The Rural Housing Assistance Grants Program consolidates funding for rural housing grant programs. This consolidation of housing grant funding provides greater flexibility to tailor financial assistance to applicant needs.

Very Low-Income Housing Repair Grants

The Very Low-Income Housing Repair Grants Program is authorized under section 504 of title V of the Housing Act of 1949. The rural housing repair grant program is carried out by making grants to very low-income families to make necessary repairs to their homes in order to make such dwellings safe and sanitary, and remove hazards to the health of the occupants, their families, or the community.

These grants may be made to cover the cost of improvements or additions, such as repairing roofs, providing toilet facilities, providing a convenient and sanitary water supply, supplying screens, repairing or providing structural supports or making similar repairs, additions, or improvements, including all preliminary and installation costs in obtaining central water and sewer service. A grant can be made in combination with a section 504 very low-in-come housing repair loan.

No assistance can be extended to any one individual in the form of a loan, grant, or combined loans and grants in excess of $27,500, and grant assistance is limited to persons, or families headed by persons who are 62 years of age or older.

Supervisory and Technical Assistance Grants

Supervisory and technical assistance grants are made to public and private non-profit organizations for packaging loan applications for housing assistance under sections 502, 504, 514/516, 515, and 533 of the Housing Act of 1949. The assistance is directed to very low-income families in underserved areas where at least 20 percent of the population is below the poverty level and at least 10 percent or more of the population resides in substandard housing. In fiscal year 1994 a Homebuyer Education Program was implemented under this authority. This program provides low-income individuals and families education and counseling on obtaining and/or maintaining occupancy of adequate housing and supervised credit assistance to become successful homeowners.

Compensation for Construction Defects

Compensation for construction defects provides funds for grants to eligible section 502 borrowers to correct structural defects, or to pay claims of owners arising from such defects on a newly constructed dwelling purchased with RHS financial assistance. Claims are not paid until provisions under the builder's warranty have been fully pursued. Requests for compensation for construction defects must be made by the owner of the property within 18 months after the date financial assistance was granted.

Rural Housing Preservation Grants

Rural housing preservation grants (section 533) of the Housing and Urban-Rural Recovery Act of 1983 (42 U.S.C. 1490m) authorizes the Rural Housing Service to administer a program of home repair directed at low- and very low-income people.

Committee Recommendations

The Committee recommends an appropriation of $40,000,000 for the Rural Housing Assistance Grants Program.

The following table compares the grant program levels recommended by the Committee to the fiscal year 2018 levels and the budget request:

Rural Housing Assistance Grants

[In thousands of dollars]			
	Fiscal year 2018 enacted	**Fiscal year 2019 budget request**	**Committee recommendation**
Very low-income housing repair grants	30,000		30,000
Housing preservation grants	10,000		10,000
Total	40,000		40,000

The Committee recommends that the Rural Housing Service prioritize funding for communities with unique weather patterns in need of replacing antiquated heating systems with more efficient technologies.

Rural Community Facilities Program Account

(Including Transfers of Funds)	
Appropriations, 2018	$48,627,000
Budget estimate, 2019	
Committee recommendation	52,063,000

Community facility loans were created by the Rural Development Act of 1972 (7 U.S.C. 1926 et seq.) to finance a variety of rural community facilities. Loans are made to organizations, including certain Indian tribes and corporations not operated for profit and public and quasi-public agencies, to construct, enlarge, extend, or otherwise improve community facilities providing essential services to rural residents. Such facilities include those providing or supporting overall community development, such as fire and rescue services, healthcare, transportation, traffic control, and community, social, cultural, and recreational benefits. Loans are made for facilities which primarily serve rural residents of open country and rural towns and villages of not more than 20,000 people. Healthcare, fire and rescue facilities, and educational facilities are the priorities of the program and receive the majority of available funds.

The Community Facility Grant Program authorized in the Federal Agriculture Improvement and Reform Act of 1996 (Public Law 104–127), is used in conjunction with the existing direct and guaranteed loan programs for the development of community facilities, such as hospitals, fire stations, and community centers. Grants are targeted to the lowest income communities. Communities that have lower population and income levels receive a higher cost-share contribution through these grants, to a maximum contribution of 75 percent of the cost of developing the facility.

Committee Recommendations

The Committee recommends an appropriation of $52,063,000 for the Rural Community Facilities Program Account.

The following table provides the Committee's recommendations, as compared to the fiscal year 2018 and budget request levels:

Rural Community Facilities Program Account

[In thousands of dollars]			
	Fiscal year 2018 enacted	**Fiscal year 2019 budget request**	**Committee recommendatio n**
Loan levels:			
Community facilities direct loans	2,800,000	3,500,000	3,000,000
Community facilities guaranteed loans	148,287		148,287
Total loan levels	2,948,287	3,500,000	3,148,287
Budget authority:			
Community facilities guaranteed loans	4,849		4,285
Community facilities grants	30,000		32,000
Economic initiative grants	5,778		5,778
Rural community development initiative	4,000		6,000
Tribal college grants	4,000		4,000
Total budget authority	48,627		52,063

Community Facilities Program Priorities

The Secretary is encouraged to prioritize Community Facilities program awards to applications that develop facilities to provide prevention, treatment, or recovery services for substance abuse disorder and for rural communities facing severe wildfire risk.

Rural Business—Cooperative Service

The Rural Business—Cooperative Service [RBS] was established by Public Law 103–354, Federal Crop Insurance Reform and Department of Agriculture Reorganization Act of 1994, dated October 13, 1994. Its programs were previously administered by the Rural Development Administration, the Rural Electrification Administration, and the Agricultural Cooperative Service.

Rural Business Program Account

(Including Transfers of Funds)	
Appropriations, 2018	$77,342,000
Budget estimate, 2019	
Committee recommendation	69,619,000

The Rural Business and Industry Loan Program was created by the Rural Development Act of 1972, and finances a variety of rural industrial development loans. Loans are made for rural industrialization and rural community facilities under Rural Development Act amendments to the Consolidated Farm and Rural Development Act (7 U.S.C. 1932 et seq.) authorities. Business and industrial loans are made to public, private, or cooperative organizations organized for profit, to certain Indian tribes, or to individuals for the purpose of improving, developing or financing business, industry, and employment or improving the economic and environmental climate in rural areas. Such purposes include financing business and industrial acquisition, construction, enlargement, repair or modernization, financing the purchase and development of land, easements, rights-of-way, buildings, payment of startup costs, and supplying working capital.

Rural business development grants were authorized by the Agricultural Act of 2014 and can be made to governmental and non-profit entities, and Indian tribes. Up to 10 percent of appropriated funds may be used to: identify and analyze business opportunities; identify, train, and provide technical assistance to existing or prospective rural entrepreneurs and managers; assist in the establishment of new rural businesses and the maintenance of existing businesses; conduct economic development planning, coordination and leadership development; and establish centers for training, technology, and trade. The balance of appropriated funding may be used for projects that support the development of business enterprises that finance or facilitate: the development of small and emerging private business enterprise; the establishment, expansion, and operation of rural distance learning networks; the development of rural learning programs; and the provision of technical assistance and training to rural communities for the purpose of improving passenger transportation.

Committee Recommendations

The Committee recommends an appropriation of $69,619,000 for the Rural Business Program Account.

The following table provides the Committee's recommendations, as compared to the fiscal year 2018 and budget request levels:

Rural Business Program Account

[In thousands of dollars]			
	Fiscal year 2018 enacted	**Fiscal year 2019 budget request**	**Committee recommendation**
Loan levels:			
Business and industry guaranteed loans loan levels	919,765		919,765
Budget authority:			
Business and industry guaranteed loans	37,342		21,339
Rural business development grants	34,000		40,280
DRA, NBRC, and ARC	6,000		8,000
Total budget authority	77,342		69,619

Regional Food Hubs

The Committee encourages USDA to partner with States and other interested partners to build and refurbish food hub and food distribution centers that serve rural farmers but are located in urban areas through programs like the Business and Industry guaranteed loan program.

Rural Business Development Grants

Rural coastal economies have often been economically disadvantaged by the loss of natural resource-related jobs and have been the first rural communities to feel the negative effects of a changing climate. As these rural coastal communities continue to have agriculture-related economic opportunity such as value-added seafood processing as well as new opportunities, the use of Rural Business Development grants may be prioritized in rural coastal communities to support innovation and job growth within all sectors, including for related infrastructure. Particular priority should be given in the case of public-private partnerships and cross-jurisdictional efforts.

Rural Business Program Account

The Committee recommends $500,000 for transportation technical assistance.

The Committee directs that of the $4,000,000 recommended for grants to benefit Federally Recognized Native American Tribes, $250,000 shall be used to implement an American Indian and Alaska Native passenger transportation development and assistance initiative.

Intermediary Relending Program Fund Account

(Including Transfer of Funds) [In thousands of dollars]			
	Fiscal year 2018 enacted	**Fiscal year 2019 budget request**	**Committee recommendation**
Estimated loan level	18,889		18,889
Direct loan subsidy	4,361		4,157
Administrative expenses	4,468		4,468
Total, loan subsidies and administrative expenses	8,829		8,625

The rural development intermediary relending loan program was originally authorized by the Economic Opportunity Act of 1964 (Public Law 88–452). The making of rural development loans by the Department of Agriculture was reauthorized by Public Law 113–79, the Agricultural Act of 2014.

Loans are made to intermediary borrowers (small investment groups) who in turn will reloan the funds to rural businesses, community development corporations, private nonprofit organizations, public agencies, et cetera, for the purpose of improving business, industry, community facilities, and employment opportunities and diversification of the economy in rural areas.

The Federal Credit Reform Act of 1990 established the program account. Appropriations to this account will be used to cover the lifetime subsidy costs associated with the direct loans obligated in 2019, as well as for administrative expenses.

Committee Recommendations

The Committee recommends an appropriation of $8,625,000 for the Intermediary Relending Program Fund.

Rural Economic Development Loans Program Account

[In thousands of dollars]	
	Estimated loan level
Fiscal year 2018 level	45,000
Fiscal year 2019 request	
Committee recommendation	45,000

The Rural Economic Development Loans program was established by the Reconciliation Act of December 1987 (Public Law 100–203), which amended the Rural Electrification Act of 1936 (Act of May 20, 1936), by establishing a new section 313. This section of the Rural Electrification Act (7 U.S.C. 901) established a cushion of credit payment program and created the rural economic development subaccount. The Administrator of RUS is authorized under the Act to utilize funds in this program to provide zero interest loans to electric and telecommunications borrowers for the purpose of promoting rural economic development and job creation projects, including funding for feasibility studies, startup costs, and other reasonable expenses for the purpose of fostering rural economic development.

Committee Recommendation

The Committee recommends a loan program level of $45,000,000, to be funded from earnings on the Cushion of Credit and fees on guaranteed underwriting loans made pursuant to section 313A of the Rural Electrification Act of 1936.

Rural Cooperative Development Grants

Appropriations, 2018	$27,550,000
Budget estimate, 2019	
Committee recommendation	30,050,000

Rural cooperative development grants are authorized under section 310B(e) of the Consolidated Farm and Rural Development Act, as amended. Grants are made to fund the establishment and operation of centers for rural cooperative development with their primary purpose being the improvement of economic conditions in rural areas. Grants may be made to nonprofit institutions or institutions of higher education. Grants may be used to pay up to 75 percent of the cost of the project and associated administrative costs. The applicant must contribute at least 25 percent from non-Federal sources, except 1994 institutions, which only need to pro-vide 5 percent. Grants are competitive and are awarded based on specific selection criteria.

Cooperative research agreements are authorized by 7 U.S.C. 2204b. The funds are used for cooperative research agreements, primarily with colleges and universities, on critical operational, organizational, and structural issues facing cooperatives.

Cooperative agreements are authorized under 7 U.S.C. 2201 to any qualified State departments of agriculture, university, and other State entity to conduct research that will strengthen and enhance the operations of agricultural marketing cooperatives in rural areas.

The Appropriate Technology Transfer for Rural Areas [ATTRA] program was first authorized by the Food Security Act of 1985. The program provides information and technical assistance to agricultural producers to adopt sustainable agricultural practices that are environmentally friendly and lower production costs.

Committee Recommendations

The Committee recommends an appropriation of $30,050,000 for Rural Cooperative Development Grants.

Of the funds recommended, $3,750,000 is for the Appropriate Technology Transfer for Rural Areas program, with $500,000 to be used for veterans' training in agriculture. The Committee has included language in the bill that not more than $3,000,000 shall be made available to cooperatives or associations of cooperatives whose primary focus is to provide assistance to small, minority producers.

Value Added

The Committee recommends $17,500,000 for value-added agricultural product market development grants. Of this amount, the Secretary is directed to make $2,500,000 for Agriculture Innovation Center funding available as grants to States authorized to host, and that have previously hosted a USDA Agriculture Innovation Center and where the State continues to demonstrate support, and provide non-Federal grant funding to producers developing, producing, and marketing value-added agricultural and food products.

Rural Energy for America Program

Appropriations, 2018	$293,000
Budget estimate, 2019	
Committee recommendation	338,000

The Rural Energy for America Program is authorized under section 9007 of the Farm Security and Rural Investment Act of 2002 (7 U.S.C. 8107). This program may fund energy audits, direct loans, loan guarantees, and grants to farmers, ranchers, and small rural businesses for the purchase of renewable energy systems and for energy efficiency improvements.

Committee Recommendations

The Committee recommends an appropriation of $338,000 for the Rural Energy for America Program.

The following table provides the Committee's recommendation as compared to the fiscal year 2018 and budget request levels:

Rural Energy for America Program

[In thousands of dollars]	Fiscal year 2018 enacted	Fiscal year 2019 budget request	Committee recommendation
Estimated loan level	7,576		7,576
Guaranteed loan subsidy	293		338

Biogas System Development

The Secretary is encouraged, in coordination with other Federal agencies, to support biogas system development with financial and technical assistance through existing energy programs, and to prioritize the collection and analysis of related environmental, technical, and economic performance data.

Energy Efficiency Coordination

The Committee directs increased coordination and cooperation among USDA agencies and offices to better utilize the energy efficiency and renewable energy programs available through the Rural Energy for America program. Additionally, no later than 120 days after enactment of this Act, USDA is directed to submit a report to the Committee detailing how the agencies make information about its energy programs accessible to rural communities and how funds are being leveraged for energy efficiency investments in rural areas.

Energy Storage Technologies

The Secretary is directed to clarify that energy storage technologies are eligible for Rural Energy for America program funding.

Rural Utilities Service

The Rural Utilities Service [RUS] was established under the Federal Crop Insurance Reform and Department of Agriculture Reorganization Act of 1994 (Public Law 103–354), October 13, 1994. RUS administers the electric and telephone programs of the former Rural Electrification Administration and the water and waste programs of the former Rural Development Administration.

Rural Water and Waste Disposal Program Account

(Including Transfers of Funds)	
Appropriations, 2018	$560,263,000
Budget estimate, 2019	
Committee recommendation	558,183,000

The water and waste disposal program is authorized by sections 306, 306A, 309A, 306C, 306D, 306E, and 310B of the Consolidated Farm and Rural Development Act (7 U.S.C. 1921 et seq., as amended). This program makes loans for water and waste development costs. Development loans are made to associations, including corporations operating on a nonprofit basis, municipalities and similar organizations, generally designated as public or quasi-public agencies, that propose projects for the development, storage, treatment, purification, and distribution of domestic water or the collection, treatment, or disposal of waste in rural areas. Such grants may not exceed 75 percent of the development cost of the projects and can supplement other funds borrowed or furnished by applicants to pay development costs.

The solid waste grant program is authorized under section 310B(b) of the Consolidated Farm and Rural Development Act. Grants are made to public bodies and private nonprofit organizations to provide technical assistance to local and regional governments for the purpose of reducing or eliminating pollution of water resources and for improving the planning and management of solid waste disposal facilities.

Committee Recommendations

The Committee recommends an appropriation of $558,183,000 for the Rural Water and Waste Disposal Program Account.

The Committee recommends $68,000,000 for water and waste disposal systems grants for Native Americans, including Native Alaskans, and the Colonias. The Committee recognizes the special needs and problems for delivery of basic services to these populations, and encourages the Secretary to distribute these funds in line with the fiscal year 2014 distribution, to the degree practicable. In addition, the Committee makes up to $19,000,000 available for the circuit rider program.

The Committee is concerned about raw sewage discharge in some rural communities in the mid-south, particularly historically impoverished communities that have had difficulty utilizing Rural Development programs. The Committee is aware of the unique challenges faced by these communities and directs the Department to develop a pilot program to coordinate with a regional university to solve untreated raw sewage issues with innovative technologies and strategic management and regulatory models. The pilot should address rural wastewater management including: county needs assessments, testing wastewater options, defining funding mechanisms for remediation and developing regulatory guidance.

The following table provides the Committee's recommendations, as compared to the fiscal year 2018 and budget request levels:

Rural Water and Waste Disposal Program Account

[In thousands of dollars]			
	Fiscal year 2018 enacted	**Fiscal year 2019 budget request**	**Committee recommendation**
Loan levels:			
Water and waste disposal direct loans	1,200,000	1,200,000	1,200,000
Water and waste disposal guaranteed loans	50,000		50,000
Total loan levels	1,250,000	1,200,000	1,250,000
Budget authority:			
Water and waste disposal direct loans	2,040		
Water and waste disposal guaranteed loans	230		190
Water and waste disposal grants	400,000		400,000
Solid waste management grants	4,000		4,000
Water well systems grants	993		993
Colonias, AK and Native American grants	68,000		68,000
Water and waste water revolving funds	1,000		1,000
High energy cost grants	10,000		10,000
Circuit rider	19,000		19,000
Emergency community water assistance grants	15,000		15,000
Technical assistance grants	40,000		40,000
Total, budget authority	560,263		558,183

Rural Electrification and Telecommunications Loans Program Account (Including Transfers of Funds)

The Rural Electrification Act of 1936 (7 U.S.C. 901 et seq.) provides the statutory authority for the electric and telecommunications programs.

The Federal Credit Reform Act of 1990 (Public Law 101–508) established the program account. An appropriation to this account will be used to cover the lifetime subsidy costs associated with the direct loans obligated and loan guarantees committed in fiscal year 2019, as well as for administrative expenses.

Committee Recommendations

The following table reflects the Committee's recommendation for the Rural Electrification and Telecommunications Loans Program Account, the loan subsidy and administrative expenses, as compared to the fiscal year 2018 and budget request levels:

Rural Electrification and Telecommunications Loans Program Account

[In thousands of dollars]			
	Fiscal year 2018 enacted	**Fiscal year 2019 budget request**	**Committee recommendation**
Loan authorizations:			
Electric:			
Direct FFB	5,500,000	5,500,000	5,500,000
Guaranteed underwriting	750,000		750,000
Telecommunications:			
Direct, Treasury Rate	345,000	172,600	345,000
Direct, FFB	345,000	517,400	345,000
Total loan authorization	6,940,000	6,190,000	6,940,000
Total budget authority	863	863	1,725

Cyber and Electric Grid Security

The Secretary is encouraged to prioritize Rural Utilities Service electric program loan awards for cyber and electric grid security improvements.

Energy Efficiency and Conservation Loan Program Oversight

The Committee is concerned that the Energy Efficiency and Conservation Loan Program does not yet have performance measures in place, as highlighted by a 2016 Inspector General report. The Committee directs USDA to institute such measures prior to awarding any additional funding through the program, or no later than 180 days from enactment of this Act.

Distance Learning, Telemedicine, and Broadband Program

[In thousands of dollars]			
	Fiscal year 2018 enacted	**Fiscal year 2019 budget request**	**Committee recommendation**
Loan and grant levels:			
Distance learning and Telemedicine Program:			
Grants	32,000	23,600	33,000
Broadband program:			
Treasury rate loans	29,851	23,149	29,851
Treasury rate loans budget authority	5,000	4,521	5,830
Grants	30,000	30,000	30,000
Total DLT and Broadband Program level	91,851	76,749	92,851
Total DLT and Broadband Budget authority	67,000	58,121	68,830

The Distance Learning, Telemedicine, and Broadband Program is authorized by the Food, Agriculture, Conservation and Trade Act of 1990 (7 U.S.C. 950aaa et seq.), as amended by the Federal Agriculture Improvement and Reform Act of 1996 (Public Law 104–127). This program provides incentives to improve the quality of phone services, to provide access to advanced telecommunications services and computer networks, and to improve rural opportunities.

This program provides the facilities and equipment to link rural education and medical facilities with more urban centers and other facilities providing rural residents access to better healthcare through technology and increasing educational opportunities for rural students. These funds are available for loans and grants.

Committee Recommendations

The Committee recommends an appropriation of $68,830,000 for the Distance Learning, Telemedicine, and Broadband Program. Funds recommended for the RUS broadband program are intended to promote broadband availability in those areas where there is not otherwise a business case for private investment in a broadband network. The Committee encourages RUS to focus expenditures on projects that bring broadband service to currently unserved households.

The Committee recommendation includes $3,000,000 to address critical healthcare needs, as authorized by section 379G of the Consolidated Farm and Rural Development Act.

Broadband Grants

Of the funds recommended, $30,000,000 in grants shall be made available to support broadband transmission for rural areas.

Broadband Infrastructure

The Committee recognizes the importance of access to high speed broadband to promote job creation and income generation in remote rural areas. For these reasons, a broadband infrastructure initiative was begun in fiscal year 2018. The Committee recommendation provides additional funding for this initiative, and encourages the Department to implement the pilot program as promptly as is feasible. The Committee notes that it is the policy of RUS to avoid duplication of efforts when financing telecommunications infrastructure via its programs; however, the Committee is concerned that this policy does not extend to the deployment of broadband-capable infrastructure. In order to prevent duplication of services, the Committee directs RUS to implement operational changes and report back to the Committee on administrative efforts to eliminate duplicative or over building of broadband technology.

Determination of Sufficient Access to Broadband

This Committee has provided substantial funding to extend high speed broadband access to remote rural areas. The Secretary was directed to reevaluate, on an annual basis, the criteria for broadband sufficient access. The Secretary is encouraged to coordinate with the Federal Communications Commission and other relevant Federal entities when making determinations of sufficient access, to ensure the most accurate and up-to-date broadband coverage data are used, while being cognizant of potential problems of overbuilding.

Locating Existing Infrastructure

The Secretary is encouraged to utilize appropriate grant program funds to locate buried, antiquated infrastructure facilities prior to construction of new utilities infrastructure financed by the Rural Utilities Service.

Multi-Strand Fiber Optic Cable

The Committee recommends that, within these funds, the Secretary explore a pilot grant program to demonstrate the use of multi-strand fiber optic cable that exist as part of electrical transmission infrastructure to provide state-of-the-art broadband services to currently underserved rural schools and medical centers within a mile of the existing cable.

RUS Grants and Loans for Open Access Infrastructure Projects

The Committee is aware that public entities have invested in open access fiber infrastructure that is facilitating the delivery of high-speed broadband services by licensed telecommunications providers, including the model pioneered by public port authorities. The Committee understands that while particular open access fiber projects may be eligible for RUS grants and loans, more generally, there exist significant barriers to government backing for these types of open access investments. The Committee believes RUS programs should support financially-feasible open access infrastructure projects that meet program goals. The Committee urges RUS to ensure the agency's criteria and application processes provide for fair consideration of open access projects by accounting for the unique structures and opportunities such projects present in advancing broadband deployment in unserved and underserved communities.

TITLE IV: DOMESTIC FOOD PROGRAMS

Office of the Under Secretary for Food, Nutrition, and Consumer Services

Appropriations, 2018	$800,000
Budget estimate, 2019	800,000
Committee recommendation	800,000

The Office of the Under Secretary for Food, Nutrition and Consumer Services provides direction and coordination in carrying out the laws enacted by the Congress with respect to the Department's nutrition assistance activities. The Office has oversight and management responsibilities for the Food and Nutrition Service.

Committee Recommendations

The Committee recommends an appropriation of $800,000 for the Office of the Under Secretary for Food, Nutrition and Consumer Services.

Food Security in Frontier Communities

The Committee appreciates the intent of the Food and Nutrition Service to focus on implementing locally-designed initiatives to increase food security in frontier communities within its area of responsibility. Helping these communities to adapt to changing growing conditions and subsistence food availability and to develop the capacity to grow more food locally will improve their tenuous food security and provide opportunities for economic development in extremely low-in-come regions. The Committee therefore strongly encourages the Food and

Nutrition Service to finalize plans to work with relevant stakeholders to develop and implement the plans that were initiated in the past year.

Nutrition Program Efficiency

The Committee encourages the Secretary to focus process and technology improvement grants within the Food and Nutrition Service [FNS] to expand public-private partnerships to increase food security in a cost-efficient and accountable manner.

Food and Nutrition Service

The Food and Nutrition Service represents an organizational effort to eliminate hunger and malnutrition in this country. Nutrition assistance programs provide access to a nutritionally adequate diet for families and persons with low incomes and encourage better eating patterns among the Nation's children. These programs include:

Child Nutrition Programs

The National School Lunch and School Breakfast, Summer Food Service, and Child and Adult Care Food programs provide funding to the States, Puerto Rico, the Virgin Islands, American Samoa, and Guam for use in serving nutritious lunches and breakfasts to children attending schools of high school grades and under, to children of preschool age in child care centers, and to children in other institutions in order to improve the health and well-being of the Nation's children, and broaden the markets for agricultural food commodities. Through the Special Milk Program, assistance is provided to the States for making reimbursement payments to eligible schools and child care institutions which institute or expand milk service in order to increase the consumption of fluid milk by children. Funds for this program are provided by direct appropriation and transfer from section 32.

Special Supplemental Nutrition Program for Women, Infants, and Children [WIC]

This program safeguards the health of pregnant, postpartum, and breast-feeding women, infants, and children up to age 5 who are at nutritional risk because of inadequate nutrition and income by providing supplemental foods. The delivery of supplemental foods may be done through health clinics, vouchers redeemable at retail food stores, or other approved methods which a cooperating State health agency may select. Funds for this program are provided by direct appropriation.

Supplemental Nutrition Assistance Program

This program seeks to improve nutritional standards of needy persons and families. Assistance is provided to eligible households to enable them to obtain a better diet by increasing their food purchasing capability, usually by furnishing benefits in the form of electronic access to funds. The program also includes Nutrition Assistance to Puerto Rico.

The program also includes the Food Distribution Program on Indian Reservations, which provides nutritious agricultural commodities to low-income persons living on or near Indian reservations who choose not to participate in the Supplemental Nutrition Assistance Program.

Commodity Assistance Program [CAP]

This program provides funding for the Commodity Supplemental Food Program [CSFP], the Farmers' Market Nutrition Program, Disaster Assistance, Pacific Island Assistance, and administrative expenses for TEFAP.

CSFP provides supplemental foods to low-income elderly persons age 60 and over.

TEFAP provides commodities and grant funds to State agencies to assist in the cost of storage and distribution of donated commodities.

Nutritious agricultural commodities are provided to residents of the Federated States of Micronesia and the Marshall Islands. Cash assistance is provided to distributing agencies to assist them in meeting administrative expenses incurred. It also provides funding for use in non-presidentially declared disasters, and for FNS' administrative costs in connection with relief for all disasters. Funds for this program are provided by direct appropriation.

Nutrition Programs Administration

Most salaries and Federal operating expenses of the Food and Nutrition Service are funded from this account. Also included is the Center for Nutrition Policy and Promotion [CNPP] which oversees improvements in and revisions to the food guidance systems, and serves as the focal point for advancing and coordinating nutrition promotion and education policy to improve the health of all Americans.

Child Nutrition Programs

(Including Transfers of Funds)	
Appropriations, 2018	$24,254,139,000
Budget estimate, 2019	23,146,940,000
Committee recommendation	23,184,012,000

The Child Nutrition Programs, authorized by the Richard B. Russell National School Lunch Act (Public Law 79–396) and the Child Nutrition Act of 1966 (Public Law 89–642), provide Federal assistance to State agencies in the form of cash and commodities for use in preparing and serving nutritious meals to children while they are attending school, residing in service institutions, or participating in other organized activities away from home. The purpose of these programs is to help maintain the health and proper physical development of America's children. Milk is provided to children either free or at a low cost, depending on their family income level. FNS provides cash subsidies to States for administering the programs and directly administers the program in the States which choose not to do so. Grants are also made for nutritional training and surveys and for State administrative expenses. Under current law, most of these payments are made on the basis of reimbursement rates established by law and applied to lunches and breakfasts actually served by the States. The reimbursement rates are adjusted annually to reflect changes in the Consumer Price Index for food away from home.

Committee Recommendations

The Committee recommends $23,184,012,000 for the Child Nutrition Programs.

Administrative Reviews

The Committee understands the importance of the formal administrative reviews state agencies conduct as required by the Healthy, Hunger-Free Kids Act. However, the Committee encourages the Secretary to return to the 5-year inspection cycle for schools that consistently comply with Federal standards to allow state agencies more flexibility in performing their oversight and on-sight technical assistance roles. High-risk schools that do not consistently comply with Federal regulations should continue to be reviewed on a more frequent basis. The Committee also encourages FNS to assist state agencies in collaborating with one another when identifying risk factors to ensure that the administrative review process is effective and consistent nationwide.

Buy American

The Committee remains supportive of existing laws requiring school food authorities to purchase domestic commodities or products to serve in school meal programs. The Committee recognizes that despite this statutory requirement, there has been an alarming increase of foreign products served in our schools. The Committee encourages the Secretary to fully define and enforce all applicable Buy American provisions within the Secretary's jurisdiction. Further, the Secretary shall report on all actions taken to comply with this directive within 180 days of enactment of this Act.

Farm to School Program

Successful implementation of Farm to School programs requires broad-based knowledge of best practices regarding coordination among farmers, processors, distributers, students, teachers, dietary and food preparation staff, and USDA professionals. Of the grant funds provided, the Committee directs the Secretary to use at least $150,000 to coordinate with established entities, such as regional Farm to School institutes, for the creation and dissemination of information on farm to school program development, and to provide practitioner education and training, and ongoing school year coaching and technical assistance.

Innovation in School Meals

The Committee is aware that there are many new, innovative, and healthy products available that meet the National School Lunch Program and School Breakfast Program nutrition standards. The Committee is concerned about FNS' interpretation of current policies that does not allow schools to get credit for serving any such product unless it visibly represents the food component in its natural or recognizable form. The Committee encourages the Secretary to allow innovative food products made from fruits, vegetables, or legumes that meet nutrition standards for school feeding programs. The Committee understands that many of these foods are already in the retail market and encourages FNS to educate children about the many ways these nutritious foods can be served and enjoyed.

Pulse Crops

The Committee recognizes the nutritional value of pulse crops for children and encourages FNS to support school food authorities in sourcing and serving pulse crops.

Summer Food Service Program

The Committee recognizes that in many rural and frontier areas of the country where homes are widely scattered, children and youth are unable to access congregate feeding sites that participate in the Summer Food

Service Program and that existing mobile food delivery efforts are not able to meet the need. The Committee supports the Food and Nutrition Service allowing State Agencies to enable Summer Food Service Program service institutions that serve such areas where eligible children and youth have barriers to access or limited access to a congregate feeding site to use their customary reimbursement payments to develop and implement innovative methods to deliver or otherwise make available foods to eligible children and youth by non-congregate means or in non-congregate settings. In addition, the Committee requests USDA submit a report within 1 year of enactment describing how many Summer Food Service Program grantees, in which states, put in place innovative methods of food delivery by non-congregate means and in non-congregate settings, what innovative methods were used, and how many additional youth were served as a result.

Vegetables in the School Breakfast Program

Current regulations regarding the substitution of starchy and non-starchy vegetables for fruit in the School Breakfast Program are creating undue burdens for school food authorities. To encourage vegetable consumption at breakfast, the Committee encourages FNS to allow any variety of vegetable to be substituted for fruit in the School Breakfast Program.

Whole Grain Waivers

Total Obligational Authority

The Committee encourages FNS to simplify the process for School Food Authorities applying for a whole grain waiver to make the process faster and more user-friendly.

The Committee's recommendation provides for the following annual rates for the child nutrition programs.

The Committee expects FNS to utilize the National Food Service Management Institute to carry out the food safety education program.

[In thousands of dollars]		
Child nutrition programs	**Fiscal year 2019 budget request**	**Committee recommendation**
School Lunch Program	11,713,000	11,713,000
School Breakfast Program	5,081,770	5,081,770
Child and Adult Care Food Program	3,933,393	3,933,393
Summer Food Service Program	519,461	519,461
Special Milk Program	8,777	8,777
State Administrative Expenses	302,906	302,906
Commodity Procurement	1,473,874	1,473,874
Team Nutrition/HUSSC/CMS	15,475	17,004
Food Safety Education	2,929	2,929
Coordinated Review.	10,000	10,000
Computer Support	12,124	12,124
Training and Technical Assistance	13,935	13,935
CNP Studies and Evaluation	21,639	21,639
Farm to School Team	3,497	3,997
Payment Accuracy	11,203	11,203
School Meal Equipment Grants		30,000
Summer EBT Demonstration	22,957	28,000

Special Supplemental Nutrition Program for Women, Infants, and Children [WIC]

Appropriations, 2018	$6,175,000,000
Budget estimate, 2019	5,750,000,000
Committee recommendation	6,150,000,000

The special supplemental nutrition program for women, infants, and children [WIC] is authorized by section 17 of the Child Nutrition Act of 1966. Its purpose is to safeguard the health of pregnant, breast-feeding and postpartum women and infants, and children up to age 5 who are at nutritional risk because of inadequate nutrition and inadequate income.

The WIC program food packages are designed to provide foods which studies have demonstrated are lacking in the diets of the WIC program target population. The authorized supplemental foods are iron-fortified breakfast cereal, fruit or vegetable juice which contains vitamin C, dry beans, peas, and peanut butter.

There are three general types of delivery systems for WIC foods: retail purchase in which participants obtain supplemental foods through retail stores; (2) home delivery systems in which food is delivered to the participant's home; and (3) direct distribution systems in which participants pick up food from a distribution outlet. The food is free of charge to all participants.

Committee Recommendations

The Committee recommends an appropriation of $6,150,000,000 for the Special Supplemental Food Program for Women, Infants, and Children [WIC].

The Committee recommendation fully funds estimated WIC participation in fiscal year 2019. The Committee recommendation includes $60,000,000 for breastfeeding support initiatives and $19,000,000 for infrastructure. The Committee recognizes new technologies, including telemedicine, that support breastfeeding mothers through access to professional breastfeeding and nutrition consultants. The Committee provides $5,000,000 for competitive grants to allow breastfeeding mothers the ability to interact with International Board Certified Lactation Consultants and all participants access to Registered Dietitians or WIC nutritionists, consistent with the goal of WIC to promote breastfeeding and nutritional health.

WIC Food Package

The Committee appreciates the work of the National Academies of Science to review and make recommendations for updating the WIC food packages to reflect current science and cultural factors. The Committee notes, however, that while all revised packages now allow some fish, the amounts remain low compared to the recommendations of authoritative agencies such as the World Health Organization and in some cases, sporadic. The Committee strongly encourages the Department to prioritize the health and cultural benefits of fish consumption as regulations are revised to implement the NAS recommendations and to increase the amount of healthful fish above the amounts recommended by the NAS. The Committee also strongly encourages the Department to allow States to prioritize fish over legumes and peanut butter to respond to the cultural preferences of WIC participants in States like Alaska.

Supplemental Nutrition Assistance Program

Appropriations, 2018	$74,013,499,000
Budget estimate, 2019	73,218,276,000
Committee recommendation	73,219,274,000

The Food Stamp Program was reauthorized through fiscal year 2012 and renamed the Supplemental Nutrition Assistance Program [SNAP] in the Food, Conservation, and Energy Act of 2008. The Supplemental Nutrition Assistance Program attempts to alleviate hunger and malnutrition among low-income persons by increasing their food purchasing power. Eligible households receive SNAP benefits with which they can purchase food through regular retail stores.

Other programs funded through SNAP include Nutrition Assistance to Puerto Rico and American Samoa, the Food Distribution Program on Indian Reservations, the Emergency Food Assistance Program, and the Community Food Projects program.

The Supplemental Nutrition Assistance Program is currently in operation in all 50 States, the District of Columbia, the Virgin Islands, and Guam. Participating households receive food benefits, the value of which is determined by household size and income. The cost of the benefits is paid by the Federal Government. As required by law, the Food and Nutrition Service annually revises household benefit allotments to reflect changes in the cost of the thrifty food plan.

Administrative Costs

All direct and indirect administrative costs incurred for certification of households, issuance of benefits, quality control, outreach, and fair hearing efforts are shared by the Federal Government and the States on a 50–50 basis.

State Antifraud Activities

Under the provisions of the Food and Nutrition Act of 2008, States are eligible to be reimbursed for 50 percent of the costs of their fraud investigations and prosecutions.

States are required to implement an employment and training program for the purpose of assisting members of households participating in SNAP in gaining skills, training, or experience that will increase their ability to obtain regular employment. The Department of Agriculture has implemented a grant program to States to assist them in providing employment and training services.

Committee Recommendations

The Committee recommends $73,219,274,000 for the Supplemental Nutrition Assistance Program. Of the amount recommended, $3,000,000,000 is made available as a contingency reserve.

Food Distribution Program on Indian Reservations [FDPIR]

Food Package

The Committee commends the Department for convening the FDPIR Food Package Review Work Group, which includes tribal representatives and staff from FNS, to increase the amount and variety of traditional foods included in FDPIR food packages and to increase the amount of foods purchased from American Indian and Alaska Native producers and businesses. The Committee directs the Department to provide a report detailing its plans to include a greater variety of traditional foods as regular components of FDPIR food baskets; its plans to identify additional Native

American and Alaska Native producers of traditional foods, including wild salmon, caribou, reindeer, elk, and other foods; and its plans to purchase additional traditional foods from a greater number of indigenous producers and businesses.

SNAP Farmers Markets

The Committee is concerned that there are unnecessary barriers and added costs for organizations that manage farmers markets in multiple locations. The Secretary shall permit such organizations to become a SNAP-authorized retailer at the level of the organization, provided that the organization notifies FNS of all market locations at which it will accept SNAP benefits. The SNAP-authorized organization will continue to bear legal responsibility for SNAP compliance at all locations it oversees, including exercising proper oversight of SNAP implementation at each participating market location.

SNAP Fraud

A January 2017 OIG report entitled "Detecting Potential SNAP Trafficking Using Data Analysis" found that FNS lacked methods to reconcile data discrepancies across their administration systems, and that retailers were providing benefits to individuals using fraudulent credentials. The Committee directs FNS to provide an update on the implementation of controls to address these problems, as well as data demonstrating whether the controls have reduced error rates.

State SNAP Implementation

The Committee is concerned about implementation of the SNAP program in certain states where states are failing to meet the required deadlines for processing applications. USDA is encouraged to work closely with States to remedy program deficiency and be aggressive in combating any falsification of SNAP implementation data.

Commodity Assistance Program

Appropriations, 2018	$322,139,000
Budget estimate, 2019	55,471,000
Committee recommendation	322,139,000

The Commodity Assistance Program includes funding for the Commodity Supplemental Food Program and funding to pay expenses associated with the storage and distribution of commodities through The Emergency Food Assistance Program.

The Commodity Supplemental Food Program [CSFP]

Authorized by section 4(a) of the Agricultural and Consumer Protection Act of 1973 (7 U.S.C. 612c note), as amended in 1981 by Public Law 97–98 and in 2014 by Public Law 113–79, this program provides supplemental food to low-income senior citizens and in some cases low-income infants and children up to age six, low-income pregnant and postpartum women. The Agricultural Act of 2014 discontinued the admission of new pregnant and postpartum women and children into the program. Those already in the program can continue to receive assistance until they are no longer eligible.

The foods for CSFP are provided by the Department of Agriculture for distribution through State agencies. The authorized commodities include: iron-fortified infant formula, rice cereal, cheese, canned juice, evaporated milk and/or nonfat dry milk, canned vegetables or fruits, canned meat or poultry, egg mix, dehydrated potatoes, farina, and peanut butter and dry beans. Elderly participants may receive all commodities except iron-fortified infant formula and rice cereal.

The Emergency Food Assistance Program [TEFAP]

Authorized by the Emergency Food Assistance Act of 1983 (7 U.S.C. 7501 et seq.), as amended, the program provides nutrition assistance to low-income people through prepared meals served on site and through the distribution of commodities to low-income households for home consumption. The commodities are provided by USDA to State agencies for distribution through State-established networks. State agencies make the commodities available to local organizations, such as soup kitchens, food pantries, food banks, and community action agencies, for their use in providing nutrition assistance to those in need.

Funds are administered by FNS through grants to State agencies which operate commodity distribution programs. Allocation of the funds to States is based on a formula which considers the States' unemployment rate and the number of persons with income below the poverty level.

Farmers' Market Nutrition Program

The Farmers' Market Nutrition Program [FMNP] provides WIC or WIC-eligible participants with coupons to purchase fresh, nutritious, unprepared foods, such as fruits and vegetables, from farmers' markets. This benefits both participants and local farmers by increasing the awareness and use of farmers' markets by low-income households.

Pacific Island and Disaster Assistance

This program provides funding for assistance to the nuclear-affected islands in the form of commodities and administrative funds. It also provides funding for use in non-presidentially declared disasters and for FNS' administrative costs in connection with relief for all disasters.

Committee Recommendations

The Committee recommends an appropriation of $322,139,000 for the Commodity Assistance Program. The Committee continues to encourage the Department to distribute Commodity Assistance Program funds equitably among the States, based on an assessment of the needs and priorities of each State and the State's preference to receive commodity allocations through each of the programs funded under this account.

Commodity Supplemental Food Program

The Committee recommends $238,120,000 for the Commodity Supplemental Food Program. This amount fully funds participation in fiscal year 2019.

Farmers' Market Nutrition Program

The Committee is aware that the Farmers' Market Nutrition Program provides fresh fruits and vegetables to low-income mothers and children, benefiting not only WIC participants, but local farmers as well. Therefore, the Committee recommends $18,548,000 for the Farmers' Market Nutrition Program and directs the Secretary to obligate these funds within 45 days.

The Emergency Food Assistance Program

The Agricultural Act of 2014 provides $294,000,000 for TEFAP commodities to be purchased with Supplemental Nutrition Assistance Program funds. The Committee recommendation includes $64,401,000 for TEFAP transportation, storage, and program integrity. In addition, the Committee recommendation grants the Secretary authority to transfer up to an additional 10 percent from TEFAP commodities for this purpose and urges the Secretary to use this authority.

The Committee encourages the Secretary to identify opportunities for increasing the supply of TEFAP commodities in the coming fiscal year through bonus and specialty crop purchases. The Department shall make available to the States domestically produced catfish fillets for distribution to local agencies.

Nutrition Programs Administration

Appropriations, 2018	$153,841,000
Budget estimate, 2019	160,838,000
Committee recommendation	164,688,000

The Nutrition Programs Administration appropriation provides for most of the Federal operating expenses of the Food and Nutrition Service, which includes the Child Nutrition Programs; Special Milk Program; Special Supplemental Nutrition Program for Women, Infants, and Children [WIC]; Supplemental Nutrition Assistance Program; Nutrition Assistance for Puerto Rico; the Commodity Assistance Program, including the Commodity Supplemental Food Program and the Emergency Food Assistance Program; and Farmers' Market Nutrition Program and Pacific Island and Disaster Assistance.

The major objective of Nutrition Programs Administration is to efficiently and effectively carry out the nutrition assistance programs mandated by law. This is to be accomplished by the following: (1) giving clear and consistent guidance and supervision to State agencies and other cooperators; (2) assisting the States and other cooperators by providing program, managerial, financial, and other advice and expertise; (3) measuring, reviewing, and analyzing the progress being made toward achieving program objectives; and (4) carrying out regular staff support functions.

Committee Recommendations

The Committee recommends an appropriation of $164,688,000 for Nutrition Programs Administration. The Committee recommendation includes $12,297,000 for the development and dissemination of the 2020 version of the Dietary Guidelines for Americans.

TITLE V: FOREIGN ASSISTANCE AND RELATED PROGRAMS

Office of the Under Secretary for Trade and Foreign Agricultural Affairs

Appropriations, 2018	$875,000
Budget estimate, 2019	875,000
Committee recommendation	875,000

The Office of the Under Secretary for Trade and Foreign Agricultural Affairs provides direction and coordination in carrying out the laws enacted by the Congress with respect to the Department's international affairs (except for foreign economic development). The Office has oversight and management responsibilities for the Foreign Agricultural Service.

Committee Recommendations

The Committee recommends an appropriations of $875,000 for the Office of the Under Secretary for Trade and Foreign Agricultural Affairs.

Food Chain Systems

The Committee is aware that the lack of comprehensive cold food chain systems is one of the main causes of food loss and results in a significant percentage of food spoilage from farm-to-market. Preventing food loss and implementing a robust cold food chain results in substantial benefits such as increased nutrition, a safer food supply, greater economic opportunity, and increased resilience. In order to maximize the benefit investment in the agricultural productivity of the developing world, the Committee encourages the Department to give strong consideration to the use of cold chain technologies and include the development of appropriate cooling technologies in programs, policies, and strategic plans aimed at hunger prevention and food security in developing agricultural markets.

Office of Codex Alimentarius

Appropriations, 2018	$3,796,000
Budget estimate, 2019	3,796,000
Committee recommendation	3,976,000

The Committee recommends an appropriation of $3,976,000 for the Office of Codex Alimentarius. Funding was previously provided through the Food Safety and Inspection Service.

Foreign Agricultural Service: Salaries and Expenses

(Including Transfers of Funds) [In thousands of dollars]			
	Appropriations	**Transfers from loan accounts**	**Total**
Appropriations, 2018	199,666	6,382	206,048
Budget estimate, 2019	193,085	6,382	199,467
Committee recommendation	212,230	6,382	218,612

The Foreign Agricultural Service [FAS] was established March 10, 1953, by Secretary's Memorandum No. 1320, supplement 1. Public Law

83–690, approved August 28, 1954, transferred the agricultural attachés from the Department of State to the Foreign Agricultural Service.

The mission of FAS overseas is to represent U.S. agricultural interests, to promote export of domestic farm products, improve world trade conditions, and report on agricultural production and trade in foreign countries. FAS staff are stationed at 98 offices around the world where they provide expertise in agricultural economics and marketing, as well as provide attaché services.

FAS carries out several export assistance programs to counter the adverse effects of unfair trade practices by competitors on U.S. agricultural trade. The Market Access Program [MAP] conducts both generic and brand-identified promotional programs in conjunction with nonprofit agricultural associations and private firms financed through reimbursable CCC payments.

The General Sales Manager was established pursuant to section 5(f) of the charter of the Commodity Credit Corporation and 15 U.S.C. 714–714p. The funds allocated to the General Sales Manager are used for conducting the following programs: (1) CCC Export Credit Guarantee Program (GSM–102), including facilities financing guarantees; (2) Food for Peace; (3) section 416b Overseas Donations Program; (4) Market Access Program; and (5) programs authorized by the Commodity Credit Corporation Charter Act including barter, export sales of most CCC-owned commodities, export payments, and other programs as assigned to encourage and enhance the export of U.S. agricultural commodities.

Committee Recommendations

The Committee recommends $218,612,000 for the Foreign Agricultural Service, including a direct appropriation of $212,230,000.

The Committee recommendation includes $3,187,000 for Capital Security Cost Sharing; $1,537,000 for International Cooperative Administrative Support Services; and $10,000,000 for trade activities.

Borlaug Fellows Program

The Committee recommendation includes $3,500,000 for the Borlaug International Agricultural Science and Technology Fellows Program. This program provides training for international scientists and policymakers from selected developing countries. The fellows work closely with U.S. specialists in their fields of expertise and apply that knowledge in their home countries. The Committee recognizes the importance of this program in helping developing countries strengthen their agricultural practices and food security.

Cochran Fellowship Program

The Committee recommendation includes $6,500,000 for the Cochran Fellowship Program. The Committee encourages the Secretary to continue to provide additional support for the program through the Commodity Credit Corporation Emerging Markets Program.

Foreign Market Development Cooperator Program

The Committee expects the FAS to fund the Foreign Market Development Cooperator Program.

Market Access Program

The Committee continues the full mandatory funding for the Market Access Program and expects the Department to administer the program as authorized in 7 U.S.C. 5623, without changing the eligibility requirements for participation of cooperative organizations, small businesses, trade associations, and other entities.

Food for Peace Title I Direct Credit and Food for Progress Program Account

(Including Transfer of Funds)	
Appropriations, 2018	$149,000
Budget estimate, 2019	142,000
Committee recommendation	142,000

Committee Recommendations

The Committee recommends an appropriation of $142,000 for administrative expenses to continue servicing existing Food for Peace title I agreements.

Food for Peace Title II Grants

Appropriations, 2018	$1,600,000,000
Budget estimate, 2019	
Committee recommendation	1,716,000,000

Commodities Supplied in Connection with Dispositions Abroad (Title II) (7 U.S.C. 1721–1726)

Commodities are supplied without cost through foreign governments to combat malnutrition and to meet famine and other emergency requirements. Commodities are also supplied for nonemergencies through public and private agencies, including intergovernmental organizations. The Commodity Credit Corporation pays ocean freight on shipments under this title, and may also pay overland transportation costs to a land-locked country, as well as internal distribution costs in emergency situations. The funds appropriated for title II are made available to private voluntary organizations and cooperatives to assist these organizations in meeting administrative and related costs.

Committee Recommendations

The Committee recommends an appropriation of $1,716,000,000 for Food for Peace title II grants.

McGovern-Dole International Food for Education and Child Nutrition Program Grants

Appropriations, 2018	$207,626,000
Budget estimate, 2019	
Committee recommendation	210,255,000

The McGovern-Dole International Food for Education and Child Nutrition Program helps support education, child development, and food security for some of the world's poorest children. The program provides for donations of U.S. agricultural products, as well as financial and technical assistance, for school feeding and maternal and child nutrition projects in low-income, food-deficit countries that are committed to universal education. Commodities made available for donation through agreements with private voluntary organizations, cooperatives, intergovernmental organizations, and foreign governments may be donated for direct feeding or for local sale to generate proceeds to support school feeding and nutrition projects.

Committee Recommendations

The Committee recommends an appropriation of $210,255,000 for the McGovern-Dole International Food for Education and Child Nutrition Program.

Local and Regional Procurement

The Committee provides an appropriation of $15,000,000 for efforts to build long-term agriculture sustainability and establish a local investment in school feeding programs. With direct U.S. commodity contributions, projects supported by the McGovern-Dole Food for Education Program have significantly improved the attendance, nourishment, and learning capacity of school-aged children in low-income countries throughout the impoverished world. New funding authorities would enable school feeding programs to proactively transition from direct commodity assistance to locally sourced agriculture products. The Committee directs the Secretary to conduct the Local and Regional Food Aid Procurement Project Program in accordance with the priorities of the McGovern-Dole International Food for Education and Child Nutrition Program.

Commodity Credit Corporation Export (Loans) Credit Guarantee Program Account

(Including Transfers of Funds) [In thousands of dollars]		
	Guaranteed loan levels	**Administrative expenses**
Appropriations, 2018	5,500,000	8,845
Budget estimate, 2019	5,500,000	6,717
Committee recommendation	5,500,000	8,845

In 1980, the Commodity Credit Corporation [CCC] instituted the Export Credit Guarantee Program (GSM–102) under its charter authority. With this program, CCC guarantees, for a fee, payments due U.S. exporters under deferred payment sales contracts (up to 36 months) for defaults due to commercial as well as noncommercial risks. The risk to CCC extends from the date of export to the end of the deferred payment period covered in the export sales contract and covers only that portion of the payments agreed to in the assurance agreement. Operation of this program is based on criteria which will assure that it is used only where it is determined that it will develop new market opportunities and maintain and expand existing world markets for U.S. agricultural commodities. The program encourages U.S. financial institutions to provide financing to those areas where the institutions would be unwilling to provide financing in the absence of the CCC guarantees. CCC also provides facilities financing guarantees.

The Federal Credit Reform Act of 1990 establishes the program account. The subsidy costs of the CCC export guarantee programs are exempt from the requirement of advance appropriations of budget authority according to section 504(c)(2) of the Federal Credit Reform Act of 1990, Public Law 101–508. Appropriations to this account will be used for administrative expenses.

TITLE VI: RELATED AGENCY AND FOOD AND DRUG ADMINISTRATION

Department of Health and Human Services

Food and Drug Administration

The Food and Drug Administration [FDA] is a scientific regulatory agency whose mission is to promote and protect the public health and safety of Americans. FDA's work is a blend of science and law. The Food and Drug Administration Amendments Act of 2007 [FDAAA] (Public Law 110–85) reaffirmed the responsibilities of the FDA: to ensure safe and effective products reach the market in a timely way, and to monitor products for continued safety while they are in use. In addition, FDA is entrusted with two critical functions in the Nation's war on terrorism: preventing willful contamination of all regulated products, including food, and improving the availability of medications to prevent or treat injuries caused by biological, chemical, radiological, or nuclear agents.

The FDA Foods program has the primary responsibility for assuring that the food supply, quality of foods, food ingredients and dietary supplements are safe, sanitary, nutritious, wholesome, and honestly labeled, and that cosmetic products are safe and properly labeled. The variety and complexity of the food supply has grown dramatically while new and more complex safety issues, such as emerging microbial pathogens, natural toxins, and technological innovations in production and processing, have developed. This program plays a major role in keeping the United States food supply among the safest in the world.

In January 2011, the Food Safety Modernization Act was signed into law. This law enables FDA to better protect public health by strengthening the food safety system. It enables FDA to focus more on preventing food safety and feed problems rather than relying primarily on reacting to problems after they occur. The law also provides FDA with new enforcement authorities designed to achieve higher rates of compliance with prevention- and risk-based food and feed safety standards and to better respond to and contain problems when they do occur. The law also gives FDA important new tools to hold imported food and feed to the same standards as domestic food and feed and directs FDA to build an integrated national food safety system in partnership with State and local authorities.

The FDA Drugs programs are comprised of four separate areas, Human Drugs, Animal Drugs, Medical Devices and Biologics. FDA is responsible for the lifecycle of products, including premarket review and postmarket surveillance of human and animal drugs, medical devices and biological products to ensure their safety and effectiveness. For Human Drugs this includes assuring that all drug products used for the prevention, diagnosis and treatment of disease are safe and effective. Additional procedures include the review of investigational new drug applications; evaluation of market applications for new and generic drugs, labeling and composition of prescription and over-the-counter drugs; monitoring the quality and safety of products manufactured in, or imported into, the United States; and, regulating the advertising and promotion of prescription drugs. The Animal Drugs and Feeds Program ensures only safe and effective veterinary drugs, intended for the treatment and/or prevention of diseases in animals and the improved production of food-producing animals, are approved for marketing.

The FDA Biologics program assures that blood and blood products, blood test kits, vaccines, and therapeutics are pure, potent, safe, effective, and properly labeled. The program inspects blood banks and blood processors, licenses and inspects firms collecting human source plasma, evaluates and licenses biologics manufacturing firms and products; lot releases licensed products; and monitors adverse events associated with vaccine immunization, blood products, and other biologics.

The FDA Devices and Radiological program ensures the safety and effectiveness of medical devices and eliminates unnecessary human exposure to manmade radiation from medical, occupational, and consumer products. In addition, the program enforces quality standards under the Mammography Quality Standards Act (Public Law 108–365). Medical devices include thousands of products from thermometers and contact lenses to heart pacemakers, hearing aids, and MRIs. Radiological products include items such as micro-wave ovens and video display terminals.

FDA's National Center for Toxicological Research in Jefferson, Arkansas, serves as a specialized resource, conducting peer-review scientific research that provides the basis for FDA to make sound science-based regulatory decisions through its premarket review and postmarket surveillance. The research is designed to define and understand the biological mechanisms of action underlying the toxicity of products and lead to developing methods to improve assessment of human exposure, susceptibility and risk of those products regulated by FDA.

In 2009, Congress granted FDA new authority to regulate the manufacture, distribution, and marketing of tobacco products. FDA exercises this responsibility by protecting the public health from the health effects of tobacco, setting scientific standards and standards for tobacco product review, conducting compliance activities to enforce its authority over tobacco, and conducting public education and outreach about the health effects of tobacco products.

Salaries and Expenses

[In thousands of dollars]			
	Appropriation	**User fees**	**Total**
Appropriations, 2018	2,800,078	2,337,963	5,138,041
Budget estimate, 2019	3,171,920	2,460,221	5,632,141
Committee recommendation	2,959,078	2,460,221	5,419,299

Committee Recommendations

The Committee recommends an appropriation of $2,959,078,000 for FDA salaries and expenses. The Committee also recommends $2,460,221,000 in definite user fees, including: $960,568,000 in Prescription Drug User Fee Act user fee collections; $196,668,000 in Medical Device User Fee and Modernization Act user fee collections; and $30,331,000 in Animal Drug User Fee Act user fee collections; $18,336,000 in Animal Generic Drug User Fee Act user fee collections; $712,000,000 in Tobacco Product user fee collections; $501,396,000 in Generic Drug User Fee Act user fee collections; and $40,922,000 in Biosimilar User Fee Act user fee collections. The Committee also recommends $58,346,000 in permanent, indefinite user fees, including: $5,300,000 in Voluntary Qualified Importer Program collections; $1,434,000 in food and feed recall collections; $6,414,000 in food reinspection collections; $20,522,000 in Mammography Quality Standards Act fee collections; $10,062,000 in color certification collections; $7,686,000 in Pediatric Disease Priority Review Voucher collections; $712,000 in third-party auditor collections; $1,520,000 in outsourcing facility collections; and $4,696,000 in export and certification fees, as assumed in the President's budget. The Committee recommendation includes bill language which prohibits FDA from developing, establishing, or operating any program of user fees authorized by 31 U.S.C. 9701.

The Committee recommendation does not include proposed user fees for food facility registration and inspection, food import, food contact substance notification, cosmetics, and international courier imports. None of these user fee proposals have been authorized by Congress. The Committee will continue to monitor any action by the appropriate authorizing Committees regarding these proposed user fees.

The Committee expects FDA to continue all projects, activities, laboratories, and programs as included in fiscal year 2018 unless otherwise specified. The Committee does not support $29,400,000 of the proposed reductions; however it does accept the $2,500,000 reduction for compounding (which was intended for one-time use) and the $1,500,000 reduction for consumer education and outreach regarding agricultural biotechnology.

The Committee recommendation includes an increase of $163,000,000 for medical product and food safety activities requested in the budget. Included in this funding is $5,000,000 to fully fund the Oncology Center of Excellence; $37,600,000 to modernize the generic drug review process; $20,000,000 for investment and innovation for rare diseases; $11,700,000 to promote domestic manufacturing; $8,200,000 for a new platform for drug development; $6,000,000 for MedTech manufacturing, $7,200,000 for FSMA cooperative agreements; $2,800,000 for food import safety; $5,000,000 to address food safety outbreaks; $500,000 to test antibiotic resistance in imported seafood; and $59,000,000 for opioid prevention activities.

The following table reflects the Committee's recommendations, as compared to the fiscal year 2018 and budget request levels:

Food and Drug Administration Salaries and Expenses

[In thousands of dollars]			
	Fiscal year 2018 enacted	**Fiscal year 2019 budget request**	**Committee recommendation**
Centers and related field activities:			
Foods	1,041,615	1,029,863	1,052,315
Center for Food Safety and Applied Nutrition [CFSAN]	316,606	315,494	320,106
Field Activities	725,009	714,369	732,209
Human Drugs	495,603	686,364	555,403
Center for Drug Evaluation and Research [CDER]	359,396	548,388	419,196
Field Activities	136,207	137,976	136,207
Biologics	215,443	251,854	218,443
Center for Biologics Evaluation and Research [CBER]	174,052	210,755	177,052
Field Activities	41,391	41,099	41,391
Animal Drugs	172,552	180,284	173,052

(Continued)

[In thousands of dollars]			
	Fiscal year 2018 enacted	**Fiscal year 2019 budget request**	**Committee recommendation**
Center for Veterinary Medicine [CVM]	107,905	115,673	108,405
Field Activities	64,647	64,611	64,647
Medical and radiological devices	330,064	455,442	336,064
Center for Devices and Radiological Health	246,319	372,588	252,319
Field Activities	83,745	82,854	83,745
National Center for Toxicological Research	63,331	65,200	65,531
Other Activities	196,275	198,565	273,075
Rent and related activities	114,987	135,927	114,987
Rental Payments to GSA	170,208	168,421	170,208
Total, FDA salaries and expenses, new budget authority	2,800,078	3,171,920	2,959,078

Added Sugar Labeling

The Committee remains concerned about potential consumer confusion over the new FDA nutritional labeling requirements for added sugar for single ingredient products like maple syrup and honey, where sugar is naturally occurring in the product rather than added to the product. The Committee is aware that the FDA has had discussions with maple and honey producers regarding their concerns that the labeling requirement as currently drafted could mislead consumers to think that sugar has been added to a pure single-ingredient maple or honey product. The Committee directs the FDA to continue working with the pure maple syrup and honey industries to ensure appropriate labeling for those single-ingredient products where sugar is naturally occurring in the product rather than added to the product. The Committee is aware of a proposed alternative Nutrition Facts labeling approach that would clearly delineate between "Added Sugars" and "Naturally Occurring Sugars" in a product, as well as another proposed alternative that would permit labeling denoting "No

Added Sugars" on applicable products, and directs the FDA to evaluate such proposals. The Committee further directs the FDA to submit a report within 60 days of enactment of this Act describing the research that was conducted prior to issuance of a final rule updating the Nutrition Facts label for packaged foods on May 27, 2016, to determine consumer perception regarding mandatory "added sugar" labeling on single ingredient products in which no sugar is added during processing, including pure maple syrup and honey.

ADUFA Reporting

The Committee supports the collection and reporting of accurate and validated data of antimicrobial drug use for food-producing animals, but is concerned that antimicrobial sales and distribution data currently reported under ADUFA 105 have been equated with actual antimicrobial use data. In order for ADUFA reporting to promote public understanding in a meaningful and accurate way, FDA should ensure that such reporting clearly describe the limitations of sales data, including that they do not represent actual use. Therefore, the Committee encourages the Agency to seek alternative methods to better identify and reduce inappropriate antimicrobial drug uses.

Alzheimer's Drug Development

The Committee applauds FDA for revising guidance on Alzheimer's drug development that supports the use of an endpoint on cognition as a biomarker for Alzheimer's disease clinical trials. We are optimistic that this approach will help advance Alzheimer's drug development, particularly evaluation of candidate therapies in patients who are asymptomatic. The Committee encourages FDA to continue supporting innovative approaches to Alzheimer's drug development, including biomarkers development platforms involving industry, academic researchers and patient preference tools.

Animal Feed Ingredients

The Committee is concerned with the slow pace of review and approval of ingredients for feed for animals. The Committee urges FDA to dedicate additional personnel to speed the review and approval process.

Anti-Counterfeiting Techniques

The Committee directs the FDA to report back on the benefits and costs of incorporating multi-layering and covert technologies with barcoding technology in meeting the provisions of the DQSA for pharmaceutical products within 90 days of the enactment of this Act.

Autoantibody Qualification

The appearance of certain islet autoantibodies in the serum of individuals increases the chance of developing type 1 diabetes at some point in the future. Therefore the Committee encourages the FDA to work with the Type 1 diabetes community on the assessment of potential diabetes biomarkers related to islet autoimmunity, which might help inform the design of clinical studies.

Biomanufacturing Innovation

The Committee supports FDA participation in public-private partnerships to accelerate biomanufacturing innovation and encourages FDA to provide funding for this purpose from within available resources.

Breast Density

The Committee recognizes the importance of patients receiving their own personal medical information and directs the Food and Drug Administration to ensure that mammography reports and summaries received by patients and their providers include appropriate information about breast density specified by the Secretary, including, at a minimum, the effect of breast density in masking the presence of breast cancer on a mammogram, the qualitative assessment of the provider who interpreted the mammogram, and a reminder to patients that individuals with dense breast tissue should talk with their providers if they have any questions or concerns about their summary.

Cancer Immunotherapy Clinical Trials

The Committee is aware of the remarkable promise of cancer immunotherapy and encouraged by the FDA's recent approval of new treatments that harness this approach to fighting cancer. More than 1,500 immuno-oncology clinical trials are in some stage of development. As more patients turn to immune-based treatments, and more clinical trials are conducted to evaluate them, understanding how to recognize and manage the side effects of cancer immunotherapies will become increasingly important. Currently, however, standard parameters for reporting cancer immunotherapy-related adverse events in clinical trials are lacking, and this makes comparisons and management across studies challenging. The Committee, therefore, urges the FDA to work with the research community and the pharmaceutical industry to develop standardized templates for reporting toxicities in cancer immunotherapy clinical trials.

Center for Safety and Nutrition Centers of Excellence

The Committee is aware of the important contribution of the FDA Center for Food Safety and Applied Nutrition's Centers of Excellence [COEs] program in supporting critical basic research as well as facilitating the implementation of the FDA Food Safety Modernization Act. The Committee encourages the Agency to continue to fully utilize the COEs to accomplish these goals, and instructs that it enhance its level of support for FDA Food Safety Modernization Act activities.

Clinical Trials

The Committee acknowledges the responsibilities of FDA to protect public health and advance medical innovation and encourages FDA to continue its efforts to improve the effectiveness of the clinical trial process.

The Committee is encouraged by the development of novel digital technologies to facilitate the use of virtual clinical trials that would make it easier for patients to participate in trials regardless of where they live.

Through telemedicine, connected sensors, patient engagement applications, and direct data capture tools, virtual trials are conducted geographically near the patient. Direct contact with the patient is still maintained remotely, but reducing or eliminating on-site visits has the potential to increase patient convenience and lower study costs. The Committee recommends that the FDA develop the necessary framework to advance the use of virtual trials while still maintaining quality data necessary for FDA approval. The FDA shall report to the Committee on their activities to advance digital technologies and the impact on patient access to clinical trials.

Computational Medicine

The Committee appreciates FDA's continued support for and use of modeling and simulation in clinical trials, as well as its work toward the establishment of an affiliation agreement with an academic institution with expertise in this field. This partnership will allow for the development of personalized medical interventions, optimizes the regulatory process with in silico clinical trials and bridges gaps in the current regulatory infrastructure. The Committee directs FDA to formalize this important function in improving outcomes and reducing costs inherent to drug and device discovery.

Contact Lens Safety

The Committee is aware that counterfeit versions of FDA-regulated medical devices exist and is concerned that certain foreign manufactured entrants in the marketplace are providing patients with mishandled or mislabeled versions of FDA-approved contact lenses. The Committee is also concerned about reports of online sales of counterfeit contact lenses. Therefore, the Committee directs the FDA to provide a report that includes a summary of foreign contact lens manufacturing facilities findings, domestic counterfeit contact lens retailer investigations, inspection activities, and any agency oversight and enforcement activities related to imported or reimported counterfeit contact lens meant for domestic sales.

Cotton Ginning

The Committee is concerned about the impact of the "Current Good Manufacturing Practice, Hazard Analysis, and Risk-Based Preventive Controls for Food for Animals" final rule (80 FR 56170; September 17, 2015) on the cotton industry. The Committee notes post-harvest activity of ginning cotton does not transform the resulting cottonseed into a "processed food," and thus, cottonseed should fall within the definition of a "raw agricultural commodity" for purposes of rules promulgated pursuant to the FSMA. In addition, the Committee is concerned about the rationale for the definitions of "primary production farm" and "secondary activities farm" and how these definitions factor into the determination of operations either being exempt from or covered by certain requirements of the final rule. Therefore, the Committee directs the FDA to provide outreach and technical assistance to cotton ginning operations to assist them in complying with the final rule or subsequent guidance documents.

Dietary Fiber

The Committee is concerned that the FDA has not issued final guidance regarding the definition of dietary fiber, and encourages the FDA to issue these final guidance documents and provide sufficient time for food manufacturers to comply.

Dietary Supplements

More than half of Americans take at least one dietary supplement each day, with use particularly prevalent among older persons and in children. While dietary supplements enter the market under the assumption that they are safe, the FDA has documented that some products are contaminated, either intentionally or unintentionally, with inherently unsafe ingredients, including active pharmaceutical ingredients. These products violate the Dietary Supplement Health and Education Act [DSHEA] and pose potential risks to consumers. The Committee applauds FDA's inspection of and enforcement actions against manufacturers with dietary supplement products that contain ingredients that are potentially harmful or otherwise noncompliant with the law. FDA has indicated it conducts roughly 500 inspections a year and issues approximately 70–80 warning letters on Current Good Manufacturing Practice [CGMP] violations. In order to better detect dangerous products in the market, FDA is encouraged to continue to invest resources into oversight and inspection of manufacturing plants that produce dietary supplements. The Committee has been pleased with the interagency collaboration and urges FDA to continue working with the Department of Justice to remove illegal dietary supplements from the market and directs increased resources toward enforcement of DSHEA, including inspection and enforcement activities. The Committee urges the FDA to issue guidance on new dietary ingredients [NDIs) for dietary supplements that is consistent with the DSHEA while continuing to use current statutory authorities to remove unsafe ingredients and products. The Committee further encourages the FDA to take industry standards and marketplace disruption into consideration when issuing any guidance on NDIs. In addition, the Committee directs FDA to submit, not later than 180 days after enactment of this Act, a report that includes the number of enforcement actions FDA brought against dietary supplement manufacturers and marketers; the manufacturers and marketers of products claiming to be dietary supplements; the number of dietary supplement good manufacturing practice inspections FDA conducted in 2017; the number of FTEs dedicated to dietary supplement inspections; and the number of serious adverse events that were reported to FDA from 2016 to 2017.

Digital Health Products

The Committee is encouraged by the FDA's efforts to implement section 3030 of the 21st Century Cures Act regarding low-risk medical software and launch of the Digital Health Software Precertification [Pre-Cert] Program to learn from software developers about their products and quality processes. The Committee believes that digital health technologies are extremely promising and that consumers should have assurances that the products work as claimed. The Committee is supportive of FDA efforts to increase oversight and enforcement over digital health products to assure that they are compliant with the appropriate regulatory frameworks.

Drug Shortages

The Committee acknowledges the strides that the FDA has made in reducing the amount of time for review of Generic Drug [ANDAs] applications, but remains concerned about the number of drugs in shortage that providers and patients rely on for care. The Committee encourages FDA to expand upon its current work to offer priority review to ANDAs to reduce the number and severity of drug shortages. The Committee directs the FDA to report to the Committee on how the agency is prioritizing ANDAs in order to mitigate the recent drug shortages, as well as what additional authorities the agency may need to alleviate drug shortages in the future.

Food Contact Notification User Fees

The Committee recommendation does not include proposed user fees.

Food Safety Mission

The Committee directs the FDA Foods Program to report to the Committee all activities and resources spent on nutrition-related activities for the Center for Food Safety and Applied Nutrition [CFSAN], associated field offices [ORA], and support components.

Foreign High Risk Inspections

The Committee has provided robust funding for this initiative over the last several years and directs the FDA to provide an update on these efforts, including estimated efficiencies and concerns, and plans to continue or expand this effort in the future.

FSMA Clarification for Small Farms

The Committee directs the FDA to provide further clarification to small farms on the requirements for compliance with the Food Safety Modernization Act, including information on the qualified exemptions available to small and very small farms and the actions required to achieve compliance under these exemptions. The Committee also urges the Food and Drug Administration to communicate with (including through appropriate guidance) and offer technical assistance to assist small farms with compliance.

FSMA Cooperative Agreements

The Committee is aware that some states that have entered into cooperative agreements under the State Produce Implementation Cooperative Agreement Program to provide education, outreach, and technical assistance have or are considering changing the state agency responsible for implementing these agreements. The Food and Drug Administration is directed to work with any state that designates a new implementing agency to ensure it can continue to receive funding under existing cooperative agreements without delay or loss of funding.

Glass Packaging Technologies

The Committee encourages the FDA to develop and issue draft guidance to industry to streamline chemistry, manufacturing, and control reporting requirements in order to expedite adoption and remove barriers for the use of innovative glass packaging technologies or processes with the capacity to improve product quality, reduce product recalls, reduce drug shortages, and improve public health.

Guidance for Stakeholder Input

The FDA Center for Veterinary Medicine [CVM] recently updated its list of guidance topics to include possible new topics for consideration as well as revisions to existing FDA CVM guidance documents. In addition to providing the traditional opportunities for public review and comment, the Committee encourages FDA to seek input from relevant industry stakeholders and appropriate scientific experts who can assist FDA in the development of and any revisions to guidance documents prior to a public comment period.

Human Drug Review Committee

The Committee strongly encourages the FDA to fully utilize its authorities under 18 U.S.C. 208(b)(3) to include no less than two members with an expertise in the indication for which the drug is meant to treat on each Advisory Committee when that Committee is reviewing a drug that has been designated as an Orphan Drug.

Improving Import Review

FDA shall report to the Committee how FDA is monitoring the impact of the reorganization under Program Alignment Group, and if such reorganization has improved the consistency of facility inspections and timeliness reviewing imports.

Intentional Adulteration

The Committee supports the important role of food defense plans to protect the food supply from acts intended to cause wide-scale harm to public health. The Committee encourages the FDA to work with businesses to provide clarity on food defense practices that will most effectively protect public health and to take into account appropriate food defense practices businesses have in place, including data supporting such practices.

Medical Gas Rulemaking

The Committee is pleased that the FDA has begun the process to develop separate regulations for medical gases. However, the Committee is concerned that the FDA has missed the statutory deadlines for rulemaking in section 1112 of Food and Drug Administration Safety and Innovation Act [FDASIA] and section 756 of the fiscal year 2017 Consolidated Appropriations Act. The FDA committed to complete separate regulations for medical gases in 1978 in its final rule on current good manufacturing practices, and the Committee believes that now is the appropriate time to complete that commitment for a separate section of regulations for medical gases. Therefore, the FDA shall issue final regulations required by the fiscal year 2017 Consolidated Appropriations Act no later than December 31, 2018.

Medically Necessary Foods

The Committee is aware that patients with significant medical need for physician-directed medical foods continue to face access challenges resulting from misperceptions on the part of some pharmacy benefit managers and insurance providers who are classifying these products as over-the-counter in direct contravention of established law, FDA guidance, and FAQs. These challenges continue to underscore the time-liness of clarifying the important pathway for review and oversight of quality medical foods. The Committee looks forward to working with FDA in this regard and requests feedback on how FDA can address the current access challenges as well as work to enhance this important category as it becomes an increasingly essential part of the healthcare system.

Misleading Maple Marketing

The Committee is concerned about the explosion of products marketed using the word maple and related iconography, which intentionally misleads consumers who perceive the use of the word maple and related iconography to mean that a food product contains some measurable quantity of maple syrup to flavor or sweeten the product, which consumers identify as a characterizing ingredient. The Committee directs the FDA to perform a detailed analysis of consumer perception of foods marketed with the word maple or related iconography.

Nanotechnology

The Committee recognizes the increased capabilities that FDA has developed to study environment, health, and safety of nanomaterials within FDA's Jefferson Laboratory Campus, including the National Center for Toxicological Research, and its consolidated headquarters at White Oak, Maryland. The Committee expects FDA to continue to support collaborative research with universities and industry on the toxicology of nanotechnology products and processes in accordance with the National Nanotechnology Initiative Environment, Health, and Safety Research Strategy as updated in October 2011.

National Antimicrobial Resistance Monitoring System

The Committee recommendation includes $11,300,000 for the National Antimicrobial Resistance Monitoring System. The Committee directs that no less than $500,000 shall be used to conduct one or more pilot studies to assess types and levels of antibiotic resistance in zoonotic bacteria on food products of species not currently tested by NARMS, such as imported seafood.

New Animal Drug Process

The Committee is concerned about the agency's approval process for genetically engineered animals for human consumption, particularly finfish. Thus, the Committee directs the agency to undertake a review of the process and report to Congress within 90 days of enactment of this Act on how the "New Animal Drug" process, created to approve drugs intended for use in animals, can be used as the approval process for genetically engineered animals for human consumption.

Olive Oil

Because of the substantial interest in and consumption of olive oil throughout the United States, driven in part by the significant scientifically-confirmed health benefits of these oils and the fact that the United States has become a globally-important producer of olive oils, especially extra virgin olive oil, the Committee directs the FDA to establish a separate U.S. Standard of Identity for different grades of olive oil (e.g., refined, virgin and extra virgin) and olive-pomace oils.

The Committee is particularly concerned with the number of different state standards for olive oils in the U.S. Because the health benefits of olive oil vary by grade, it is important to establish a uniform set of the standards to better inform and protect consumers. Extra virgin olive oil is the highest quality of olive oil and provides the greatest health benefits for consumers. The FDA is directed to consult and meet with domestic producers and importers of olive oil to develop a science-based Standard of Identity for extra virgin olive oil and olive oil best suited to ensure the integrity of these products for U.S. consumers.

Opioids

The Committee remains deeply concerned about the opioid abuse epidemic that has taken the lives of more than 350,000 Americans from 1999–2016, including more than 42,000 in 2016. As such, the Committee recommendation includes $59,000,000 for FDA to continue its increased activities related to the crisis. This funding is for the FDA to better identify and target firms and organizations importing illicit drugs into the U.S.; maintain increased staffing to inspect packages and the number of packages being inspected; maintain enhanced criminal investigation resources; and maintain increased staff and equipment to efficiently screen imported products.

As the agency that oversees the approval of these drugs, the FDA has a responsibility to consider the public health impact of opioid misuse, abuse, diversion, and overdose death, while considering the needs of patients with chronic conditions who require regular use of opioid therapies to properly manage their debilitating condition. The Committee supports FDA's commitment to addressing this crisis through all available authorities, and continues to encourages them to continue implementation of the Opioid Action Plan to determine how innovative changes in opioid packaging, distribution, and medication disposal procedures can help mitigate the national opioid crisis, including working to support ongoing efforts at the state and regional level.

The Committee continues its directive for FDA to refer any drug application for an opioid to an advisory committee for their recommendations prior to approval, unless the FDA finds that holding such advisory committee is not in the interest of protecting and promoting public health.

The Committee is concerned about marketed opioid products that pose disproportionate overdose risk due to their formulation in which the daily recommended dosage far exceeds the CDC's threshold for dangerous daily opioid intake. The Committee directs the FDA to examine existing data and filings from leading medical and health societies, and remove from the market any ultra-high-dose opioids that the FDA finds are unsafe and pose a public health hazard.

The Committee notes that, even with recent decreases, opioid prescribing rates dramatically exceed current standards for accepted and effective medical use, with nearly 14 billion opioid doses put on the United States market each year. Therefore, the Committee believes that it is imperative that FDA, consistent with its own Advisory Committee recommendations, take any and all steps necessary to require continuing medical education, aligned with the most recent Centers for Disease Control and Prevention's Guidelines for Prescribing Opioids for Chronic Pain, for providers who write opioid prescriptions, including through the Risk Evaluations and Mitigation Strategy.

The Committee is aware that 80 percent of individuals with opioid use disorder were introduced to opioids via prescription and notes that a growing body of evidence indicates the co-prescription of naloxone along with certain prescriptions has the potential to reduce overdose deaths and opioid overdose related healthcare costs. The Committee directs the Commissioner to seek recommendations from the Drug Safety and Risk Management Advisory Committee regarding a framework for the inclusion of information in the labeling and/or REMS of drugs that are opioids or used in Medically-Assisted Treatment relating to the co-prescription of opioid over-dose reversal drugs along with opioids prescribed to patients that meet CDC guidelines as at risk for overdose.

Oversight Activities

The Committee recommendation includes $1,500,000 for the HHS Office of Inspector General specifically for oversight of FDA activities.

Patient Experience

The Committee is aware of FDA's implementation of policies to promote public access to information about how patient experience information factored into the review of approved products. The Committee supports this step forward and encourages FDA to continue refining the instrument and ways to improve its visibility. The Committee also requests that FDA consider ways to include patient-experience information in relevant labeling and accompanying documentation to inform patient/ provider decisionmaking and payer determinations.

Patient-Focused Drug Development

The Committee is appreciative of the steps the FDA has taken to implement subtitle A of title III within the 21st Century Cures Act to better incorporate patient experience in the drug development and approval processes and requests a status report from FDA on implementation of these provisions including any challenges or impediments being faced.

Pediatric Device Consortia Grants

The Committee is pleased that the seven FDA-funded Pediatric Device Consortia have assisted in the development of more than 1,000 potential pediatric medical devices since its inception in 2009, as well as promoting job-growth in the healthcare sector, and as such, continues to support this critical effort. The program funds consortia to assist innovators in developing medical and surgical devices designed for the unique needs of children that often go unmet by devices currently available on the market. The Committee recommendation includes no less than the fiscal year 2018 funding level for Pediatric Device Consortia Grants.

Polypharmacy

The routine usage of five or more prescription medications within the same period is becoming increasingly prevalent among older adults, elevating risk factors for drug-drug interactions and adverse events. The Committee directs the FDA to assess potential impacts of polypharmacy, which might help inform the design of clinical studies.

Ready to Eat Foods

The Committee is aware that FDA is in the process of finalizing guidance regarding Listeria monocytogenes [Lm] in RTE foods. Reducing incidents of listeriosis is an important health goal and the Committee supports efforts to accomplish this objective. The Committee urges FDA to complete a comprehensive risk assessment to ensure any final guidance document is realistic and fully based in science prior to making any changes to the action level of Lm in RTE foods.

Seafood Advisory

The Committee remains concerned that the FDA published final seafood advice for pregnant and nursing women on January 18, 2017, without going through the necessary interagency review, consumer focus group testing, or the opportunity for the public to comment on the scientific peer review. Therefore, the Committee directs the FDA to reissue the final "Advice About Eating Fish" (published in 82 Fed. Reg. 6571 (January 19, 2017)) in a manner that is consistent with the FDA's nutrition science on the net effects of seafood consumption.

Shellfish Safety

The Committee urges FDA to complete the single laboratory validation of the liquid chromatography mass spectrometry [LC–MS]-based method for detecting brevetoxins associated with neurotoxic shellfish poisoning in molluscan shellfish. The Committee encourages adoption by the Interstate Shellfish Sanitation Conference of FDA's proposal for the LC–MS method for brevetoxin testing of shellfish as an Approved Method under the National Shellfish Sanitation Program.

Sunscreen Labeling Regulations

The Committee remains significantly concerned that the FDA has not approved a new over-the-counter [OTC] sunscreen ingredient since implementation of the Sunscreen Innovation Act, which improved the process by which the FDA reviews sunscreen ingredients and required the FDA to finalize an effective sunscreen monograph within 5 years. The Committee directs the FDA to meet with sponsors regarding the development of a testing regimen for sunscreen ingredients, consistent with current scientific standards, that appropriately balances the benefit of additional skin cancer prevention tools versus the risk of skin cancer. The Committee also directs FDA to maintain funding for agency efforts to clear this backlog of sunscreen applications.

In addition, the Committee is disappointed that FDA has not yet finalized a rule limiting the maximum Sun Protection Factor [SPF] to "50" or "50+" as directed by the fiscal year 2018 Consolidated Appropriations Act, and as such the Committee directs FDA to finalize the rule immediately. The Committee is also disappointed that FDA failed to issue a proposed rule to establish testing and labeling standards for sunscreen sprays and directs FDA to do so immediately.

User Fee Negotiations

The Committee affirms the important role of user fees to support programs across the FDA, and supports the negotiations between the agency and regulated industry partners to compose goals letter establishing clear expectations for both parties regarding timelines and processes associated with implementation of the law. Historically these goals letters are added to the Congressional Record, unedited by Congress, and referenced in the law authorizing the collection of such fees. The Committee is concerned that recent user fee negotiations between FDA and regulated industries have resulted in goals letters submitted to Congress containing policy changes that require statutory changes, and presume that Congress will adopt suggested statutory changes.

While the Committee encourages the agency to continue to provide suggested statutory changes in a timely manner to Congress that can help the agency meet its mission, the Committee finds that it is inappropriate for the agency and its regulated industry partners to negotiate statutory or other legal changes as part of user fee goals letters.

Vibrio

The Committee is aware of the public health challenge related to the naturally occurring bacteria called Vibrio parahaemolyticus that can accumulate in shellfish and believes that more scientific research is necessary to develop proper controls that will reduce the risk to consumers and sustain a healthy domestic shellfish industry. The Committee encourages the Food and Drug Administration [FDA] to increase funding for research into Vibrio illnesses associated with the consumption of raw molluscan shellfish, improve risk assessment models, and develop improved rapid detection methods for virulent Vibrio strains.

White Oak Expansion

The Committee is aware of the need for FDA facilities to accommodate an anticipated expanded workforce due to broader missions related to food safety and other mandates in legislation over the last few years. In the Committee's report for fiscal year 2016, the Committee requested a feasibility study to update and issue a revised Master Plan for land inside and contiguous to the White Oak Campus in order to address its expanded workforce and the facilities needed to accommodate them. The Committee directs FDA to complete this study as soon as possible. Due to the challenging fiscal environment, the Committee encourages the FDA and GSA to consider innovative financing options and partnership opportunities with non-federal government entities that provide reasonable cost options contiguous to the White Oak campus.

Youth Tobacco Use Prevention

While FDA has recently announced a new Youth Tobacco Prevention Plan to attempt to curb the use of e-cigarettes among youth, the Committee is concerned with the irresponsible marketing by some manufacturers, as well as the role characterizing flavors play in youth initiation of tobacco products. In March 2018, FDA issued an Advanced Notice of Proposed Rulemaking to examine regulatory options for tobacco product flavorings. The Committee strongly encourages the agency to complete the regulatory process in an expeditious manner, ideally within 1 year, and in a way that supports prevention of youth tobacco initiation. The agency is instructed to provide the Committee with a timeframe for when the regulatory process will be completed. Additionally, the Committee is concerned that FDA is not fully enforcing their prohibition of new or changed e-cigarettes and other nicotine products after August 8, 2016, without prior FDA review and authorization. Therefore, the Committee directs FDA to order the removal of any "deemed tobacco products" introduced after the August 8, 2016 deadline without first seeking the required FDA authorization.

Finally, the Committee is concerned about the lack of adequate age verification rules to prevent Internet sales of e-cigarette tobacco products to children, and directs FDA to establish these rules within 1 year, both at the time of sale and delivery of the product.

Buildings and Facilities

Appropriations, 2018	$11,788,000
Budget estimate, 2019	11,788,000
Committee recommendation	11,788,000

FDA maintains offices and staff in 49 States and in the District of Columbia and Puerto Rico, including field laboratories and specialized facilities, as well as the National Center for Toxicological Research complex. Repairs, modifications, improvements, and construction to FDA headquarters and field facilities must be made to preserve the properties, ensure employee safety, meet changing program requirements, and permit the agency to keep its laboratory methods up to date.

Committee Recommendations

The Committee recommends an appropriation of $11,788,000 for FDA buildings and facilities. This funding shall be used to upgrade FDA facilities and laboratories which are currently below public safety standards and incapable of performing agency requirements.

FDA Innovation Account, Cures Act

Appropriations, 2018	$60,000,000
Budget estimate, 2019	70,000,000
Committee recommendation	70,000,000

The Committee recommends $70,000,000 for the FDA as authorized in the 21st Century Cures Act.

Independent Agency: Farm Credit Administration

Limitation on Administrative Expenses

Limitation, 2018	$70,600,000
Budget estimate, 2019	74,600,000
Committee recommendation	74,600,000

The Farm Credit Administration [FCA] is the independent agency in the executive branch of the Government responsible for the examination and regulation of the banks, associations, and other institutions of the Farm Credit System.

Activities of the Farm Credit Administration include the planning and execution of examinations of Farm Credit System institutions and the preparation of examination reports. FCA also promulgates regulations, establishes standards, enforces rules and regulations, and approves certain actions of the institutions.

The administration and the institutions under its jurisdiction now operate under authorities contained in the Farm Credit Act of 1971, Public Law 92–181, effective December 10, 1971. Public Law 99–205, effective December 23, 1985, restructured FCA and gave the agency regulatory authorities and enforcement powers.

The Act provides for the farmer-owned cooperative system to make sound, adequate, and constructive credit available to farmers and ranchers and their cooperatives, rural residences, and associations and other entities upon which farming operations are dependent, and to modernize existing farm credit law to meet current and future rural credit needs.

The Agricultural Credit Act of 1987 authorized the formation of the Federal Agricultural Mortgage Corporation [FAMC] to operate a secondary market for agricultural and rural housing mortgages. The Farm Credit Administration, under section 8.11 of the Farm Credit Act of 1971, as amended, is assigned the responsibility of regulating this entity and assuring its safe and sound operation.

Expenses of the Farm Credit Administration are paid by assessments collected from the Farm Credit System institutions and by assessments to the Federal Agricultural Mortgage Corporation.

Committee Recommendations

The Committee recommends a limitation of $74,600,000 on administrative expenses of the Farm Credit Administration.

TITLE VII: GENERAL PROVISIONS (INCLUDING RESCISSIONS AND TRANSFERS OF FUNDS)

The Committee recommends the following provisions:

- Section 701. This section makes funds available for the purchase, replacement, and hire of passenger motor vehicles.
- Section 702. This section gives the Secretary of Agriculture authority to transfer unobligated balances to the Working Capital Fund and clarifies longstanding practices associated with the Fund.
- Section 703. This section limits the funding provided in the bill to 1 year, unless otherwise specified.
- Section 704. This section limits negotiated indirect costs on cooperative agreements between the Department of Agriculture and nonprofit organizations to 10 percent.
- Section 705. This section makes appropriations to the Department of Agriculture for the cost of direct guaranteed loans available until expended to disperse obligations for certain Rural Development programs.
- Section 706. This section prohibits the purchase of new information technology equipment in excess of $25,000 without the prior approval of the Chief Information Officer.

- Section 707. This section makes funds for certain conservation programs available until expended to disburse certain obligations made in the current fiscal year.
- Section 708. This section makes certain former Rural Utilities Service borrowers eligible for the Rural Economic Development loan and grant program.
- Section 709. This section provides funds for Rural Development and the Farm Service Agency information technology expenses.
- Section 710. This section includes language regarding first-class travel.
- Section 711. This section includes language regarding the Commodity Credit Corporation.
- Section 712. This section makes funds available for the expenses and activities of certain advisory committees, panels, commissions, and task forces at the Department of Agriculture.
- Section 713. This section includes language regarding the limitation on direct costs for grants awarded by the National Institute of Food and Agriculture.
- Section 714. This section includes language regarding information technology systems.
- Section 715. This section includes language regarding the availability of funds for certain Department of Agriculture programs.
- Section 716. This section prohibits the use of funds for user fee proposals that fail to provide sufficient budget impact information.
- Section 717. This section prohibits the reprogramming of funds for programs, projects, or activities in excess of $500,000 or 10 percent, whichever is less without the prior notification of the Committee on Appropriations.
- Section 718. This section includes language for the establishment of a fee under the business and industry loan program.
- Section 719. This section prohibits the Department of Agriculture and the Department of Health and Human Services from

transmitting questions or responses as a result of the appropriations hearing process to non-Department employees.

- Section 720. This section includes language regarding pre-packaged news.
- Section 721. This section requires Department of Agriculture agencies to provide reimbursement to other Department of Agriculture agencies for employees detailed for longer than 60 days.
- Section 722. This section includes language regarding farmer stress.
- Section 723. This section includes language regarding spending plans.
- Section 724. This section includes language regarding a rescission of funds.
- Section 725. This section includes language regarding section 502 single family direct loans.
- Section 726. This section includes language regarding loans and loan guarantees.
- Section 727. This section includes language regarding credit card refunds.
- Section 728. This section includes language regarding the Supplemental Nutrition Assistance Program.
- Section 729. This section includes language regarding industrial hemp.
- Section 730. This section includes language regarding the Agriculture and Food Research Initiative.
- Section 731. This section includes language regarding housing loan programs.
- Section 732. This section includes language regarding disclosure of information for pharmaceuticals.
- Section 733. This section includes language regarding gene editing.

- Section 734. This section includes language regarding dried spent grain products.
- Section 735. This section includes language regarding Geographically Disadvantaged Farmers.
- Section 736. This section includes language regarding partially hydrogenated oils.
- Section 737. This section includes language regarding the Rural Utilities Service.
- Section 738. This section includes language regarding hiring authorities.
- Section 739. This section includes language regarding the Conservation Reserve Program.
- Section 740. This section includes language regarding FDA regulation.
- Section 741. This section includes language regarding the Water Bank Program.
- Section 742. This section includes language regarding domestic preference.
- Section 743. This section includes language regarding Rural Economic Area Partnership zones.
- Section 744. This section includes language regarding Rural Development programs.
- Section 745. This section provides funding for a pilot program through the Rural Housing Service.
- Section 746. This section includes language regarding lobbying. Section 747. This section includes language regarding the Agriculture Risk Coverage program.
- Section 748. This section includes language regarding housing programs.
- Section 749. This section includes language regarding Farm to School programs.
- Section 750. This section includes language regarding sodium reduction

- Section 751. This section includes language regarding the Healthy Food Financing Initiative.
- Section 752. This section includes language regarding citrus greening.
- Section 753. This section includes language regarding FDA regulation.
- Section 754. This section includes language Distance Learning and Telemedicine programs.
- Section 755. This section includes language regarding the National Institute of Food and Agriculture.
- Section 756. This section includes language regarding rural broadband.
- Section 757. This section includes language regarding water and waste programs.

Program, Project, and Activity

During fiscal year 2019, for purposes of the Balanced Budget and Emergency Deficit Control Act of 1985 (Public Law 99–177) or the Balanced Budget and Emergency Deficit Control Reaffirmation Act of 1987 (Public Law 100–119), the following information provides the definition of the term “program, project, and activity” for departments and agencies under the jurisdiction of the Agriculture, Rural Development, Food and Drug Administration, and Related Agencies Subcommittee. The term “program, project, and activity” shall include the most specific level of budget items identified in the Agriculture, Rural Development, Food and Drug Administration, and Related Agencies Appropriations Act, 2019, and the accompanying Senate Report.

If a sequestration order is necessary, in implementing the Presidential order, departments and agencies shall apply any percentage reduction required for fiscal year 2019 pursuant to the provisions of Public Law 99–177 or Public Law 100–119 to all items specified in the explanatory notes submitted to the Committees on Appropriations of the House and Senate in support of the fiscal year 2019 budget estimates, as amended, for such departments and agencies, as modified by congressional action, and in addition:

For the Agricultural Research Service the definition shall include specific research locations as identified in the explanatory notes.

For the Natural Resources Conservation Service the definition shall include individual flood prevention projects as identified in the explanatory notes and individual operational watershed projects as summarized in the notes.

For the Farm Service Agency the definition shall include individual, regional, State, district, and county offices.

Compliance with Paragraph 7, Rule XVI of the Standing Rules of the Senate

Paragraph 7 of rule XVI requires that Committee reports accompanying general appropriations bills identify each recommended amendment which proposes an item of appropriation which is not made to carry out the provisions of an existing law, a treaty stipulation, or an act or resolution previously passed by the Senate during that session.

The Committee is filing an original bill, which is not covered under this rule, but reports this information in the spirit of full disclosure.

The Committee recommends funding for the following programs or activities which currently lack authorization for fiscal year 2019:

- Broadband Telecommunications Grants
- Child Nutrition Program State Administrative Expenses
- Farmers Market Nutrition Program
- Multi-family Housing Revitalization Program
- National School Lunch Act – Information Clearinghouse
- School Meals Program – Compliance and Accountability

- Special Supplemental Nutrition Program for Women, Infants and Children
- Supplemental Nutrition Assistance Program
- Summer Food Service Program

Compliance with Paragraph 7(C), Rule XXVI of the Standing Rules of the Senate

Pursuant to paragraph 7(c) of rule XXVI, on May 24, 2018, the Committee ordered favorably reported an original bill (S. 2976) making appropriations for Agriculture, Rural Development, Food and Drug Administration, and Related Agencies programs for the fiscal year ending September 30, 2019, and for other purposes, provided, that the bill be subject to amendment and that the bill be consistent with its budget allocation, and provided that the Chairman of the Committee or his designee be authorized to offer the substance of the original bill as a Committee amendment in the nature of a substitute to the House companion measure, by a recorded vote of 31–0, a quorum being present. The vote was as follows:

Yeas	Nays
Chairman Shelby	
Mr. McConnell	
Mr. Alexander	
Ms. Collins	
Ms. Murkowski	
Mr. Graham	
Mr. Blunt	
Mr. Moran	
Mr. Hoeven	
Mr. Boozman	
Mrs. Capito	
Mr. Lankford	

Yeas	Nays
Mr. Daines	
Mr. Kennedy	
Mr. Rubio	
Mrs. Hyde-Smith	
Mr. Leahy	
Mrs. Murray	
Mrs. Feinstein	
Mr. Durbin	
Mr. Reed	
Mr. Tester	
Mr. Udall	
Mrs. Shaheen	
Mr. Merkley	
Mr. Coons	
Mr. Schatz	
Ms. Baldwin	
Mr. Murphy	
Mr. Manchin	
Mr. Van Hollen	

Compliance with Paragraph 12, Rule XXVI of the Standing Rules of the Senate

Paragraph 12 of rule XXVI requires that Committee reports on a bill or joint resolution repealing or amending any statute or part of any statute include "(a) the text of the statute or part thereof which is proposed to be repealed; and (b) a comparative print of that part of the bill or joint resolution making the amendment and of the statute or part thereof proposed to be amended, showing by stricken-through type and italics, parallel columns, or other appropriate typographical devices the omissions and insertions which would be made by the bill or joint resolution if enacted in the form recommended by the Committee."

In compliance with this rule, changes in existing law proposed to be made by the bill are shown as follows: existing law to be omitted is enclosed in black brackets; new matter is printed in italic; and existing law in which no change is proposed is shown in roman.

TITLE 42—THE PUBLIC HEALTH AND WELFARE

CHAPTER 13—SCHOOL LUNCH PROGRAMS

1758. Program requirements

(a) Nutritional requirements

***

(h) Food safety

(1) In general

***

(3) Audits and reports by States

[For fiscal year 2018] *For fiscal year 2019*, each State shall annually—

***

(4) Audit by the Secretary

[For fiscal year 2018] *For fiscal year 2019*, the Secretary shall annually audit State reports of food safety inspections of schools submitted under paragraph (3).

***

1769g. Information clearinghouse

(a) In general

(d) Funding

Out of any moneys in the Treasury not otherwise appropriated, the Secretary of the Treasury shall pay to the Secretary to provide to the organization selected under this section, to establish and maintain the information clearinghouse, $200,000 for each of fiscal years 1995 and 1996, $150,000 for fiscal year 1997, $100,000 for fiscal year 1998, $166,000 for each of fiscal years 1999 through 2004, and $250,000 for each of fiscal years [2010 through 2018] *2010 through 2019*. The Secretary shall be entitled to receive the funds and shall accept the funds, without further appropriation.

Budgetary Impact of Bill

Prepared in Consultation with the Congressional Budget Office Pursuant to Sec. 308(A), Public Law 93–344, as Amended

[In millions of dollars]				
	Budget authority		**Outlays**	
	Committee allocation	Amount in bill	Committee allocation	Amount in bill
Comparison of amounts in the bill with the subcommittee allocation for 2019: Subcommittee on Agriculture, Rural Development, Food and Drug Administration, and Related Agencies:				
Mandatory	107,124	107,124	98,764	[1]98,764
Discretionary	23,235	23,235	24,446	[1]24,446
Security				NA
Nonsecurity	23,235	23,235	NA	NA
Projection of outlays associated with the recommendation:				
2019				[2]104,603
2020				6,170
2021				1,433
2022				743
2023 and future years				529
Financial assistance to State and local governments for				[2]33,321
2019	NA	40,173	NA	

[1]Includes outlays from prior–year budget authority.

[2]Excludes outlays from prior–year budget authority.

NA: Not applicable.

NOTE.—Pursuant to section 1002(b)(3)(B) of the 21st Century Cures Act (Public Law 114–255), $70 million in budget authority and the resulting outlays do not count for the purposes of estimates under the Congressional Budget and Impoundment Control Act of 1974 or the Balanced Budget and Emergency Deficit Control Act of 1985.

Comparative Statement of New Budget (Obligational) Authority for Fiscal Year 2018 and Budget Estimates and Amounts Recommended in the Bill for Fiscal Year 2019 (in Thousands of Dollars)

Item	2018 appropriation	Budget estimate	Committee recommendation	Senate Committee recommendation compared with (+ or -)	
				2018 appropriation	Budget estimate
TITLE I—AGRICULTURAL PROGRAMS					
Processing, Research, and Marketing					
Office of the Secretary					
Office of the Secretary	5,051	4,850	5,051		+201
Assistant to the Secretary for Rural Development	800	800	800		
Office of Homeland Security	1,496	1,448	1,496		+48
Office of Partnerships and Public Engagement	4,711	1,672	4,711		+3,039
Office of the Assistant Secretary for Administration	804	875	804		-71
Departmental Administration	22,301	22,501	22,301		-200
Subtotal, Departmental Administration	23,105	23,376	23,105		-271
Office of the Assistant Secretary for Congressional Relations	3,869	3,091	3,869		+778
Office of Communications	7,500	7,261	7,500		+239

Item	2018 appropriation	Budget estimate	Committee recommendation	Senate Committee recommendation compared with (+ or -)	
				2018 appropriation	Budget estimate
Total, Office of the Secretary	46,532	42,498	46,532		+4,034
Executive Operations:					
Office of the Chief Economist	19,786	19,487	19,786		+299
Office of Hearings and Appeals	15,222	14,183	15,222		+1,039
Office of Budget and Program Analysis	9,525	8,631	9,525		+894
Subtotal, Executive Operations	44,533	42,301	44,533		+2,232
Office of the Chief Information Officer	58,950	62,524	63,950	+5,000	+1,426
Office of the Chief Financial Officer	6,028	5,536	6,028		+492
Office of the Assistant Secretary for Civil Rights	901	800	901		+101
Office of Civil Rights	24,206	22,345	24,206		+1,861
Building and Facilities:					
Agriculture Buildings and Facilities	64,414	58,330	58,330	-6,084	
Hazardous materials management	3,503	3,463	3,503		+40
Office of Inspector General	98,208	87,436	98,208		+10,772
Office of the General Counsel	44,546	41,717	45,146	+600	+3,429
Office of Ethics	4,136	2,897	4,136		+1,239
Total, Departmental Administration	395,957	369,847	395,473	-484	+25,626

(Continued)

Item	2018 appropriation	Budget estimate	Committee recommendation	Senate Committee recommendation compared with (+ or -)	
				2018 appropriation	Budget estimate
Office of the Under Secretary for Research, Education, and Economics	800	800	800		
Economic Research Service.	86,757	45,000	86,757		+41,757
National Agricultural Statistics Service	191,717	165,000	174,767	-16,950	+9,767
Census of Agriculture	(63,350)	(45,300)	(45,300)	(-18,050)	
Agricultural Research Service:					
Salaries and expenses	1,202,766	1,018,991	1,300,966	+98,200	+281,975
Buildings and facilities	140,600			-140,600	
Total, Agricultural Research Service	1,343,366	1,018,991	1,300,966	-42,400	+281,975
National Institute of Food and Agriculture:					
Research and education activities	887,171	794,479	898,535	+11,364	+104,056
Native American Institutions Endowment Fund	(11,880)	(11,857)	(11,880)		(+23)
Extension activities	483,626	450,185	486,692	+3,066	+36,507
Integrated activities.	37,000	13,037	38,000	+1,000	+24,963
Total, National Institute of Food and Agriculture	1,407,797	1,257,701	1,423,227	+15,430	+165,526

Item	2018 appropriation	Budget estimate	Committee recommendation	Senate Committee recommendation compared with (+ or -)	
				2018 appropriation	Budget estimate
Office of the Under Secretary for Marketing and Regulatory Programs	901	800	901		+101
Animal and Plant Health Inspection Service:					
Salaries and expenses	981,893	739,151	1,000,493	+18,600	+261,342
Buildings and facilities	3,175	2,852	3,175		+323
Total, Animal and Plant Health Inspection Service	985,068	742,003	1,003,668	+18,600	+261,665
Agricultural Marketing Service:					
Marketing Services	151,595	118,617	155,845	+4,250	+37,228
Standardization activities (user fees).	(65,000)			(-65,000)	
(Limitation on administrative expenses, from fees collected)	(61,227)	(60,982)	(60,982)	(-245)	
Funds for strengthening markets, income, and supply (section 32):					
Permanent, section 32	1,344,000	1,374,000	1,374,000	+30,000	
Marketing agreements and orders (transfer from section 32)	(20,705)	(20,489)	(20,489)	(-216)	
Payments to States and Possessions.	1,235	1,109			+126
Limitation on inspection and weighing services	(55,000)	(80,000)	(55,000)	.	(-25,000)

(Continued)

Item	2018 appropriation	Budget estimate	Committee recommendation	Senate Committee recommendation compared with (+ or -)	
				2018 appropriation	Budget estimate
Total, Agricultural Marketing Service program	1,613,057	1,634,708	1,647,062	+34,005	+12,354
Office of the Under Secretary for Food Safety	800	800	800		
Food Safety and Inspection Service	1,056,844	1,032,273	1,049,344	-7,500	+17,071
Lab accreditation fees	(1,000)	(1,000)	(1,000)		
Total, Processing, Research, and Marketing	6,966,837	6,126,941	6,967,783	+946	+840,842
Total, Title I, Agricultural Programs	6,966,837	6,126,941	6,967,783	+946	+840,842
(By transfer)	(20,705)	(20,489)	(20,489)	(-216)	.
(Loan authorization)					
(Limitation on administrative expenses)	(116,227)	(140,982)	(115,982)	(-245)	(-25,000)
TITLE II—FARM PRODUCTION AND CONSERVATION PROGRAMS					
Farm Production Programs					
Office of the Under Secretary for Farm Production and Conservation	901	875	901		+26

Item	**2018 appropriation**	**Budget estimate**	**Committee recommendation**	**Senate Committee recommendation compared with (+ or -)**	
				2018 appropriation	Budget estimate
Farm Production and Conservation Business Center	1,028	196,402	1,028		-195,374
(Transfer from ACIF)		(16,081)			(-16,081)
Farm Service Agency:					
Salaries and expenses	1,202,146	920,490	1,202,146		+281,656
(Transfer from Food for Peace (Public Law 480))	(149)	(142)	(142)	(-7)	
(Transfer from export loans)	(2,463)	(335)	(2,463)		(+2,128)
(Transfer from ACIF)	(314,998)	(266,436)	(314,998)		(+48,562)
Subtotal, transfers from program accounts	(317,610)	(266,913)	(317,603)	(-7)	(+50,690)
Total, Salaries and expenses	(1,519,756)	(1,187,403)	(1,519,749)	(-7)	(+332,346)
State mediation grants	3,904	3,228	3,904		+676
Grassroots source water protection program	6,500		6,500		+6,500
Dairy indemnity program	500	500	500		
Subtotal, Farm Service Agency	1,213,050	924,218	1,213,050		+288,832
Agricultural Credit Insurance Fund [ACIF] Program Account:					
Loan authorizations:					
Farm ownership loans:					
Direct.	(1,500,000)	(1,500,000)	(1,500,000)		

(Continued)

Item	2018 appropriation	Budget estimate	Committee recommendation	Senate Committee recommendation compared with (+ or -)	
				2018 appropriation	Budget estimate
Guaranteed	(2,750,000)	(2,750,000)	(2,750,000)		
Subtotal	(4,250,000)	(4,250,000)	(4,250,000)		
Farm operating loans:					
Direct	(1,530,000)	(1,500,000)	(1,530,000)		(+30,000)
Unsubsidized guaranteed	(1,960,000)	(1,600,000)	(1,960,000)		(+360,000)
Subtotal	(3,490,000)	(3,100,000)	(3,490,000)		(+390,000)
Emergency loans	(25,610)	(37,668)	(37,668)	(+12,058)	
Indian tribe land acquisition loans	(20,000)	(20,000)	(20,000)		
Conservation loans:					
Guaranteed	(150,000)	(150,000)	(150,000)		
Indian highly fractionated land loans	(10,000)		(10,000)		(+10,000)
Boll weevil eradication loans	(60,000)	(60,000)	(60,000)		
Total, Loan authorizations	(8,005,610)	(7,617,668)	(8,017,668)	(+12,058)	(+400,000)
Loan subsidies:					
Farm operating loans:					
Direct	61,812	58,500	59,670	-2,142	+1,170
Unsubsidized guaranteed	21,756	17,280	21,168	-588	+3,888
Subtotal	83,568	75,780	80,838	-2,730	+5,058
Emergency Loans	1,260	1,567	1,567	+307	
Indian highly fractionated land loans	2,272	.	2,134	-138	+2,134

Item	2018 appropriation	Budget estimate	Committee recommendation	Senate Committee recommendation compared with (+ or -)	
				2018 appropriation	Budget estimate
Total, Loan subsidies and grants	87,100	77,347	84,539	-2,561	+7,192
ACIF administrative expenses:					
Salaries and expenses (transfer to FSA)	314,998	266,436	314,998		+48,562
Administrative expenses	10,070	10,070	10,070		
Administrative expenses (transfer to FPAC Business Center)		16,081			-16,081
Total, ACIF expenses	325,068	292,587	325,068		+32,481
Total, Agricultural Credit Insurance Fund	412,168	369,934	409,607	-2,561	+39,673
(Loan authorization)	(8,005,610)	(7,617,668)	(8,017,668)	(+12,058)	(+400,000)
Total, Farm Service Agency	1,627,147	1,491,429	1,624,586	-2,561	+133,157
Risk Management Agency (RMA):					
RMA Salaries and expenses	74,829	37,942	74,829		+36,887
Subtotal, Risk Management Agency	74,829	37,942	74,829		+36,887
Total, Farm Production Programs	1,701,976	1,529,371	1,699,415	-2,561	+170,044
Natural Resources Conservation Service:					
Private Lands Conservation Operations	874,107	669,033	879,107	+5,000	+210,074
Transfer					

(Continued)

Item	2018 appropriation	Budget estimate	Committee recommendation	Senate Committee recommendation compared with (+ or -)	
				2018 appropriation	Budget estimate
Farm Security and Rural Investment program (transfer authority)		(850,200)			(-850,200)
Total, Conservation operations	874,107	669,033	879,107	+5,000	+210,074
Watershed flood and prevention operations	150,000	.	150,000	..	+150,000
Watershed rehabilitation program	10,000			-10,000	.
Total, Natural Resources Conservation Service	1,034,107	669,033	1,029,107	-5,000	+360,074
Corporations					
Federal Crop Insurance Corporation:					
Federal crop insurance corporation fund	8,913,000	8,687,000	8,687,000	-226,000	
Commodity Credit Corporation Fund:					
Reimbursement for net realized losses	14,284,847	15,410,000	15,410,000	+1,125,153	
Hazardous waste management (limitation on expenses)	(5,000)	(5,000)	(5,000)		
Total, Corporations	23,197,847	24,097,000	24,097,000	+899,153	

Item	2018 appropriation	Budget estimate	Committee recommendation	Senate Committee recommendation compared with (+ or -)	
				2018 appropriation	Budget estimate
Total, Title II, Farm Production and Conservation Programs	25,933,930	26,295,404	26,825,522	+891,592	+530,118
TITLE III—RURAL DEVELOPMENT					
Rural Development:					
Rural development expenses:					
Salaries and expenses	230,835	156,054	232,835	+2,000	+76,781
(Transfer from RHIF)	(412,254)	(244,249)	(412,254)		(+168,005)
(Transfer from RCFP)		(147,591)			(-147,591)
(Transfer from RDLFP)	(4,468)		(4,468)		(+4,468)
(Transfer from RETLP)	(33,270)	(38,027)	(33,270)		(-4,757)
(Transfer from DLTBP)		(8,057)			(-8,057)
(Transfer from RWWDP)		(18,149)			(-18,149)
Subtotal, Transfers from program accounts	(449,992)	(456,073)	(449,992)		(-6,081)
Total, Rural development expenses	(680,827)	(612,127)	(682,827)	(+2,000)	(+70,700)
Rural Housing Service:					
Rural Housing Insurance Fund Program Account:					
Loan authorizations:					
Single family direct (Sec. 502)	(1,100,000)		(1,100,000)		(+1,100,000)
Unsubsidized guaranteed	(24,000,000)	(24,000,000)	(24,000,000)		

(Continued)

Item	**2018 appropriation**	**Budget estimate**	**Committee recommendation**	**Senate Committee recommendation compared with (+ or -)**	
				2018 appropriation	Budget estimate
Subtotal, Single family	(25,100,000)	(24,000,000)	(25,100,000)		(+1,100,000)
Housing repair (Sec. 504)	(28,000)		(28,000)		(+28,000)
Rental housing (Sec. 515)	(40,000)		(40,000)		(+40,000)
Multi–family housing guarantees (Sec. 538)	(230,000)	(250,000)	(230,000)		(-20,000)
Site development loans (Sec. 524)	(5,000)		(5,000)		(+5,000)
Single family housing credit sales	(10,000)	(10,000)	(10,000)		
Self–help housing land development housng loans (Sec. 523)	(5,000)		(5,000)		(+5,000)
Farm Labor Housing (Sec. 514)	(23,855)		(23,855)		(+23,855)
Total, Loan authorizations	(25,441,855)	(24,260,000)	(25,441,855)		(+1,181,855)
Loan subsidies:					
Single Family Direct (Sec. 502)	42,350		53,900	+11,550	+53,900
Housing repair (Sec. 504)	3,452		3,419	-33	+3,419
Rental housing (Sec. 515)	10,524		9,484	-1,040	+9,484
Farm labor housing (Sec. 514)	6,374		5,945	-429	+5,945
Self–Help Land Development Housing Loans (Sec. 523)	368		431	+63	+431
Site Development Loans (Sec. 524)	58	.	176	+118	+176
Total, Loan subsidies	63,126		73,355	+10,229	+73,355

Item	2018 appropriation	Budget estimate	Committee recommendation	Senate Committee recommendation compared with (+ or -)	
				2018 appropriation	Budget estimate
Farm labor housing grants	8,336		8,336		+8,336
RHIF administrative expenses (transfer to RD)	412,254	244,249	412,254		+168,005
Total, Rural Housing Insurance Fund program	483,716	244,249	493,945	+10,229	+249,696
(Loan authorization)	(25,441,855)	(24,260,000)	(25,441,855)	.	(+1,181,855)
Rental Assistance Program:					
Rental assistance (Sec. 521)	1,345,293	1,331,400	1,331,400	-13,893	
Multi–Family Housing Revitalization Program Account:					
Rural Housing Voucher Program	25,000	20,000	26,000	+1,000	+6,000
Multi–Family Housing Revitalization Program	22,000		24,000	+2,000	+24,000
Total, Multi–family housing revitalization	47,000	20,000	50,000	+3,000	+30,000
Mutual and self–help housing grants	30,000		30,000		+30,000
Rural housing assistance grants	40,000		40,000		+40,000
Rural Community Facilities Program Account:					
Loan authorizations:					
Community facility:					
Direct	(2,800,000)	(3,500,000)	(3,000,000)	(+200,000)	(-500,000)

(Continued)

Item	2018 appropriation	Budget estimate	Committee recommendation	Senate Committee recommendation compared with (+ or -)	
				2018 appropriation	Budget estimate
Guaranteed.	(148,287)		(148,287)	.	(+148,287)
Total, Loan authorizations	(2,948,287)	(3,500,000)	(3,148,287)	(+200,000)	(-351,713)
Loan subsidies and grants:					
Community facility:					
Guaranteed.	4,849		4,285	-564	+4,285
Grants	30,000		32,000	+2,000	+32,000
Rural community development initiative	4,000		6,000	+2,000	+6,000
Economic impact initiative grants	5,778		5,778		+5,778
Tribal college grants	4,000		4,000		+4,000
RCFP administrative expenses (transfer to RD)		147,591			-147,591
Total, RCFP Loan subsidies and grants	48,627	147,591	52,063	+3,436	-95,528
Subtotal, grants and payments	118,627	147,591	122,063	+3,436	-25,528
Total, Rural Housing Service	1,994,636	1,743,240	1,997,408	+2,772	+254,168
(Loan authorization)	(28,390,142)	(27,760,000)	(28,590,142)	(+200,000)	(+830,142)
Rural Business--Cooperative Service:					

Item	2018 appropriation	Budget estimate	Committee recommendation	Senate Committee recommendation compared with (+ or -)	
				2018 appropriation	Budget estimate
Rural Business Program Account:					
(Guaranteed business and industry loans)	(919,765)		(919,765)		(+919,765)
Loan subsidies and grants:					
Guaranteed business and industry subsidy	37,342		21,339	-16,003	+21,339
Rural business development grants	34,000		40,280	+6,280	+40,280
DRA,NBRC, and ARC	6,000		8,000	+2,000	+8,000
Total, RBP loan subsidies and grants	77,342		69,619	-7,723	+69,619
Intermediary Relending Program Fund Account:					
(Loan authorization)	(18,889)		(18,889)		(+18,889)
Loan subsidy	4,361		4,157	-204	+4,157
Administrative expenses (transfer to RD)	4,468		4,468		+4,468
Total, IRP Fund	8,829		8,625	-204	+8,625
Rural Economic Development Loans Program Account:					
(Loan authorization)	(45,000)		(45,000)		(+45,000)
Rural Cooperative Development Grants:					
Cooperative development	5,800		5,800		+5,800

(Continued)

Item	2018 appropriation	Budget estimate	Committee recommendation	Senate Committee recommendation compared with (+ or -)	
				2018 appropriation	Budget estimate
Appropriate Technology Transfer for Rural Areas	2,750		3,750	+1,000	+3,750
Grants to assist minority producers	3,000		3,000		+3,000
Value–added agricultural product market development	16,000		17,500	+1,500	+17,500
Total, Rural Cooperative Development grants	27,550		30,050	+2,500	+30,050
Rural Energy for America Program:					
(Loan authorization)	(7,576)		(7,576)		(+7,576)
Loan subsidy and grants	293		338	+45	+338
Total, Rural Energy for America Program	293		338	+45	+338
Total, Rural Business–Cooperative Service	114,014		108,632	-5,382	+108,632
(Loan authorization)	(991,230)		(991,230)		(+991,230)
Rural Utilities Service:					
Rural water and waste disposal program account:					
Loan authorizations:					
Direct	(1,200,000)	(1,200,000)	(1,200,000)		

Item	2018 appropriation	Budget estimate	Committee recommendation	Senate Committee recommendation compared with (+ or -)	
				2018 appropriation	Budget estimate
Guaranteed	(50,000)		(50,000)		(+50,000)
Total, Loan authorization	1,250,000	1,200,000	1,250,000		+50,000
Loan subsidies and grants:					
Direct subsidy.	2,040		.	-2,040	
Guaranteed subsidy	230		190	-40	+190
Water and waste revolving fund	1,000		1,000		+1,000
Water well system grants	993		993		+993
Colonias and AK/HI grants	68,000		68,000		+68,000
Water and waste technical assistance	40,000		40,000		+40,000
Circuit rider program	19,000		19,000		+19,000
Solid waste management grants	4,000		4,000		+4,000
High energy cost grants	10,000		10,000		+10,000
Water and waste disposal grants	400,000		400,000		+400,000
306A(i)(2) grants	15,000		15,000		+15,000
WWDP Administrative expenses (transfer to RD)		18,149			-18,149
Total, Loan subsidies and grants	560,263	18,149	558,183	-2,080	+540,034
Rural Electrification and Telecommunications Loans Program Account:					

(Continued)

Item	2018 appropriation	Budget estimate	Committee recommendation	Senate Committee recommendation compared with (+ or -)	
				2018 appropriation	Budget estimate
Loan authorizations:					
Electric:					
Direct, FFB	(5,500,000)	(5,500,000)	(5,500,000)		
Guaranteed underwriting	(750,000)		(750,000)		(+750,000)
Subtotal, Electric	(6,250,000)	(5,500,000)	(6,250,000)		(+750,000)
Telecommunications:					
Direct, Treasury rate	(345,000)	(172,600)	(345,000)		(+172,400)
Direct, FFB	(345,000)	(517,400)	(345,000)		(-172,400)
Subtotal, Telecommunications	(690,000)	(690,000)	(690,000)		
Total, Loan authorizations	(6,940,000)	(6,190,000)	(6,940,000)		(+750,000)
Loan Subsidy:					
Telecommunications Direct, Treasury Rate	863	863	1,725	+862	+862
Total, Loan subsidies	863	863	1,725	+862	+862
RETLP administrative expenses (transfer to RD)	33,270	38,027	33,270		-4,757
Total, Rural Electrification and Telecommunications Loans Program Account	34,133	38,890	34,995	+862	-3,895

Item	2018 appropriation	Budget estimate	Committee recommendation	Senate Committee recommendation compared with (+ or -)	
				2018 appropriation	Budget estimate
(Loan authorization)	(6,940,000)	(6,190,000)	(6,940,000)		(+750,000)
Broadband Program:					
Loan authorizations:					
Broadband telecommunications	(29,851)	(23,149)	(29,851)		(+6,702)
Total, Loan authorizations	(29,851)	(23,149)	(29,851)		(+6,702)
Loan subsidies and grants:					
Distance learning and telemedicine:					
Grants	32,000	23,600	33,000	+1,000	+9,400
Broadband telecommunications:					
Direct	5,000	4,521	5,830	+830	+1,309
Grants	30,000	30,000	30,000		
Total, Loan subsidies and grants	67,000	58,121	68,830	+1,830	+10,709
DLTBP administrative expenses (transfer to RD)		8,057			-8,057
Total, Rural Utilities Service	661,396	123,217	662,008	+612	+538,791
(Loan authorization)	(8,219,851)	(7,413,149)	(8,219,851)		(+806,702)
Total, Title III, Rural Development Programs	3,000,881	2,022,511	3,000,883	+2	+978,372
(By transfer)	(449,992)	(456,073)	(449,992)		(-6,081)
(Loan authorization)	(37,601,223)	(35,173,149)	(37,801,223)	(+200,000)	(+2,628,074)
TITLE IV—DOMESTIC FOOD PROGRAMS					

(Continued)

Item	2018 appropriation	Budget estimate	Committee recommendation	Senate Committee recommendation compared with (+ or -)	
				2018 appropriation	Budget estimate
Office of the Under Secretary for Food, Nutrition and Consumer Services	800	800	800		
Food and Nutrition Service:					
Child nutrition programs	24,196,139	23,123,983	23,126,012	-1,070,127	+2,029
School breakfast program equipment grants	30,000		30,000		+30,000
Demonstration projects (Summer EBT)	28,000	22,957	28,000		+5,043
Total, Child nutrition programs	24,254,139	23,146,940	23,184,012	-1,070,127	+37,072
Special supplemental nutrition program for women, infants, and children (WIC)	6,175,000	5,750,000	6,150,000	-25,000	+400,000
Supplemental Nutrition Assistance Program:					
(Food stamp program)	71,012,501	70,218,276	70,218,276	-794,225	
Reserve	3,000,000	3,000,000	3,000,000		
FDPIR nutrition education services	998		998		+998
Total, Food stamp program	74,013,499	73,218,276	73,219,274	-794,225	+998
Fiscal year 2018	(74,013,499)	(73,218,276)	(73,219,274)	(-794,225)	(+998)

Item	**2018 appropriation**	**Budget estimate**	**Committee recommendation**	**Senate Committee recommendation compared with (+ or -)**	
				2018 appropriation	Budget estimate
Commodity Assistance Program:					
Commodity Supplemental Food Program	238,120		238,120		+238,120
Farmers Market Nutrition Program	18,548		18,548		+18,548
Emergency Food Assistance Program	64,401	54,401	64,401		+10,000
Pacific island and disaster assistance	1,070	1,070	1,070		
Total, Commodity Assistance Program	322,139	55,471	322,139		+266,668
Nutrition Programs Administration	153,841	160,838	164,688	+10,847	+3,850
Total, Food and Nutrition Service	104,918,618	102,331,525	103,040,113	-1,878,505	+708,588
Fiscal year 2018	(104,918,618)	(102,331,525)	(103,040,113)	(-1,878,505)	(+708,588)
Total, Title IV, Domestic Food Programs	104,919,418	102,332,325	103,040,913	-1,878,505	+708,588
Fiscal year 2018	(104,918,618)	(102,331,525)	(103,040,113)	(-1,878,505)	(+708,588)
TITLE V—FOREIGN ASSISTANCE AND RELATED PROGRAMS					
Office of the Under Secretary for Trade and Foreign Agricultural Affairs	875	875	875		
Office Codex Alimentarius	3,796	3,796	3,976	+180	+180
Foreign Agricultural Service					

(Continued)

Item	**2018 appropriation**	**Budget estimate**	**Committee recommendation**	**Senate Committee recommendation compared with (+ or -)**	
				2018 appropriation	Budget estimate
Salaries and expenses	199,666	193,085	212,230	+12,564	+19,145
(Transfer from export loans)	(6,382)	(6,382)	(6,382)		
Total, Salaries and expenses	206,048	199,467	218,612	+12,564	+19,145
Food for Peace Title I Direct Credit and Food for Progress Program Account, Administrative Expenses:					
Farm Service Agency, Salaries and expenses (transfer to FSA)	149	142	142	-7	
Food for Peace Title II Grants:					
Expenses	1,600,000		1,716,000	+116,000	+1,716,000
Commodity Credit Corporation Export Loans Program Account (administrative expenses):					
Salaries and expenses (Export Loans):					
Foreign Agriculture Service, S&E (transfer to FAS)	6,382	6,382	6,382		
Farm Service Agency S&E (transfer to FSA)	2,463	335	2,463		+2,128
Total, CCC Export Loans Program Account	8,845	6,717	8,845		+2,128
McGovern–Dole International Food for Education and Child Nutrition program grants	207,626		210,255	+2,629	+210,255

Item	2018 appropriation	Budget estimate	Committee recommendation	Senate Committee recommendation compared with (+ or -)	
				2018 appropriation	Budget estimate
Total, Title V, Foreign Assistance and Related Programs	2,020,957	204,615	2,152,323	+131,366	+1,947,708
(By transfer)	(6,382)	(6,382)	(6,382)		
TITLE VI—RELATED AGENCIES AND FOOD AND DRUG ADMINISTRATION					
DEPARTMENT OF HEALTH AND HUMAN SERVICES					
Food and Drug Administration					
Salaries and expenses, direct appropriation	2,800,078	3,171,920	2,959,078	+159,000	-212,842
Prescription drug user fees	(911,346)	(960,568)	(960,568)	(+49,222)	
Medical device user fees	(193,291)	(196,668)	(196,668)	(+3,377)	
Human generic drug user fees	(493,600)	(501,396)	(501,396)	(+7,796)	
Biosimilar biological products user fees	(40,214)	(40,922)	(40,922)	(+708)	
Animal drug user fees	(18,093)	(30,331)	(30,331)	(+12,238)	
Animal generic drug user fees	(9,419)	(18,336)	(18,336)	(+8,917)	
Tobacco product user fees	(672,000)	(712,000)	(712,000)	(+40,000)	
Subtotal, user fees, enacted and definite	(2,337,963)	(2,460,221)	(2,460,221)	(+122,258)	

(Continued)

Item	2018 appropriation	Budget estimate	Committee recommendation	Senate Committee recommendation compared with (+ or -)	
				2018 appropriation	Budget estimate
Subtotal (including user fees)	(5,138,041)	(5,632,141)	(5,419,299)	(+281,258)	(-212,842)
Mammography user fees	(20,522)	(20,522)	(20,522)		
Export and color certification user fees	(14,758)	(14,758)	(14,758)		
Food and Feed Recall user fees	(1,434)	(1,434)	(1,434)		
Food Reinspection fees	(6,414)	(6,414)	(6,414)		
Voluntary qualified importer program fees	(5,300)	(5,300)	(5,300)		
Pharmacy compounding fees (CBO estimate)	(1,446)	(1,520)	(1,446)		(-74)
Priority review vouchers (PRV) pediatric disease	(7,686)	(7,686)	(7,686)		
Third party auditor	(1,400)	(712)	(712)	(-688)	
Over–the–Counter Monograph fees (legislative proposal)		(22,000)			(-22,000)
Increased export certification fees		(4,280)			(-4,280)
Subtotal, FDA user fees	(2,396,923)	(2,544,847)	(2,518,493)	(+121,570)	(-26,354)

Item	**2018 appropriation**	**Budget estimate**	**Committee recommendation**	**Senate Committee recommendation compared with (+ or -)**	
				2018 appropriation	Budget estimate
Subtotal, FDA (including user fees)	(5,197,001)	(5,716,767)	(5,477,571)	(+280,570)	(-239,196)
Buildings and facilities	11,788	11,788	11,788		
FDA Innovation account	60,000	70,000	70,000	+10,000	
Offset of appropriation pursuant to Section 1002(b)(3)(B) of the 21st Century Cures Act (Public Law114–255)	-60,000	-70,000	-70,000	-10,000	
Total, FDA (w/user fees, including proposals)	(5,208,789)	(5,728,555)	(5,489,359)	(+280,570)	(-239,196)
Total, FDA (w/enacted user fees only)	(5,208,789)	(5,728,555)	(5,489,359)	(+280,570)	(-239,196)
Total, FDA (excluding user fees)	2,811,866	3,183,708	2,970,866	+159,000	-212,842
INDEPENDENT AGENCIES					
Farm Credit Administration (limitation on administrative expenses)	(70,600)	(74,600)	(74,600)	(+4,000)	..
Total, Title VI, Related Agencies and Food and Drug Administration	2,811,866	3,183,708	2,970,866	+159,000	-212,842
TITLE VII—GENERAL PROVISIONS					
Limit Dam Rehab (Sec. 714(1))		-46,150			+46,150
Limit Environmental Quality Incentives Program (Sec. 714(2))		-136,260			+136,260

(Continued)

Item	2018 appropriation	Budget estimate	Committee recommendation	Senate Committee recommendation compared with (+ or -)	
				2018 appropriation	Budget estimate
Limit Rural Energy for America Program (Sec. 714 (2)) (rescission)					
Limit Biomass Crop Assistance Program (Sec. 714(3))	-21,000			+21,000	
Limit Biorefinery Assistance (Sec. 714(4))					
Limit Ag Management Assistance (Sec. 714 (5))		-9,380			+9,380
Limit Biorefinery Assistance (Sec. 714 (4)) (cancellation)					
Section 32 (Sec. 715) (rescission)		-342,000			+342,000
APHIS B&F—Fruit Fly Rearing (Sec. 743)					
WIC (rescission)	-800,000	-215,000	-400,000	+400,000	-185,000
Limit CSP (Sec. 730)		-27,000			+27,000
Rural Water Waste Disposal (Sec. 726)		-51,000			+51,000
RCFP (Sec. 727)		-3,046			+3,046
FAS S&E (Sec. 729)		-18,000			+18,000
TEFAP (Sec. 748)					

Item	2018 appropriation	Budget estimate	Committee recommendation	Senate Committee recommendation compared with (+ or -)	
				2018 appropriation	Budget estimate
Water and Waste	500,000		400,000	-100,000	+400,000
Emergency livestock assistance program (Sec. 756)					
Citrus Greening	7,500		7,500		+7,500
NIFA, Research & Education					
RD balances (Sec. 758) (rescission)					
Healthy Food Financing Initiative	1,000		1,000		+1,000
RD unobligated balances (rescission)					
ARS B&F unobligated balances (rescission)					
Hardwood Trees (Reforestation Pilot Program)	600			-600	
Water Bank program	4,000		4,000		+4,000
Geographic Disadvantaged farmers	1,996		1,996		+1,996
Emergency Conservation Program					
Food for Peace	116,000			-116,000	
Rural Energy Savings Program	8,000		10,000	+2,000	+10,000
Maturing mortgage pilot	1,000		1,000		+1,000
FSA ARC pilot	5,000		5,000		+5,000
NIFA Military Veteran Grants	5,000		5,000		+5,000
Conservation Reserve Program Pilot	1,000		1,000		+1,000

(Continued)

Item	2018 appropriation	Budget estimate	Committee recommendation	Senate Committee recommendation compared with (+ or -)	
				2018 appropriation	Budget estimate
Child Nutrition Training pilot	2,000			-2,000	
Broadband Pilot	600,000		425,000	-175,000	+425,000
Opioid Enforcement and Surveillance	94,000			-94,000	
Electric Loan Refinancing	5,000			-5,000	
Distance Learning Telemedicine	20,000		20,000		+20,000
Farm to School	5,000		5,000		+5,000
NIFA Leasing	6,000			-6,000	
Tree Assistance Program	15,000			-15,000	
Farmer Stress			2,000	+2,000	+2,000
Total, Title VII, General Provisions	577,096	-847,836	488,496	-88,600	+1,336,332
OTHER APPROPRIATIONS					
SUPPLEMENTAL APPROPRIATIONS FOR DISASTER RELIEF REQUIREMENTS, 2017					
Office of the Secretary (emergency)	2,360,000			-2,360,000	
Office of Inspector General (emergency)	3,000			-3,000	

Item	2018 appropriation	Budget estimate	Committee recommendation	Senate Committee recommendation compared with (+ or -)	
				2018 appropriation	Budget estimate
ARS, Buildings and Facilities (emergency)	22,000			-22,000	
Emergency Conservation Program (emergency)	400,000			-400,000	
Watershed and Flood and Prevention Operations (emergency)	541,000			-541,000	
Rural Housing Insurance Fund Program Account (emergency).	19,000			-19,000	
Rural Water and Waste Disposal Program Account (emergency)	165,000			-165,000	
Special Supplemental Nutrition Program for Women, Infants, and Children (WIC) (emergency)	14,000			-14,000	
Commodity Assistance Program (emergency)	24,000			-24,000	
FDA, Buildings and Facilities (emergency)	8,000			-8,000	
Sec. 20101(a) LIP (emergency).	3,000			-3,000	
Sec. 20101(b) ELAP (emergency)	30,000			-30,000	
Sec. 20101(c) TAP (emergency)	7,000			-7,000	
Sec. 20101(d) payment limitations (emergency).	2,000			-2,000	

(Continued)

Item	2018 appropriation	Budget estimate	Committee recommendation	Senate Committee recommendation compared with (+ or -)	
				2018 appropriation	Budget estimate
Sec. 60101 Cotton/Dariy (emergency)	47,000			-47,000	
Total, Supplemental Appropriations for Disaster Relief Requirements (Public Law 115–56)	3,645,000			-3,645,000	
ADDITIONAL SUPPLEMENTAL APPROPRIATIONS FOR DISASTER RELIEF REQUIREMENTS ACT, 2017					
Supplemental Nutritional Assistance Program (emergency)	1,270,000			-1,270,000	
Total, Other Appropriations	4,915,000			-4,915,000	
(Emergency)	4,915,000			-4,915,000	
Grand total	151,145,985	139,317,668	145,446,786	-5,699,199	+6,129,118
Appropriations fiscal year 2018	(147,030,985)	(139,874,668)	(145,846,786)	(-1,184,199)	(+5,972,118)
Emergency appropriations	(4,915,000)			(-4,915,000)	.
Disaster relief					
Rescissions	(-800,000)	(-557,000)	(-400,000)	(+400,000)	(+157,000)
Advance appropriations, fiscal year 2019					
(By transfer)	(794,689)	(765,938)	(794,466)	(-223)	(+28,528)
(Loan authorization)	(45,606,833)	(42,790,817)	(45,818,891)	(+212,058)	(+3,028,074)
(Limitation on administrative expenses)	(191,827)	(220,582)	(195,582)	(+3,755)	(-25,000)

In: Agricultural Policy …
Editor: Albertus Brinkhuis
ISBN: 978-1-53615-215-9

Chapter 2

AGRICULTURE AND RELATED AGENCIES: FY2019 APPROPRIATIONS*

Jim Monke

SUMMARY

The Agriculture appropriations bill funds the U.S. Department of Agriculture (USDA) except for the Forest Service. It also funds the Food and Drug Administration (FDA) and—in even-numbered fiscal years—the Commodity Futures Trading Commission (CFTC).

Agriculture appropriations include both mandatory and discretionary spending. Discretionary amounts, though, are the primary focus during the bill's development. The largest discretionary spending items are the Special Supplemental Nutrition Program for Women, Infants, and Children (WIC); agricultural research; FDA; rural development; foreign food aid and trade; farm assistance programs; food safety inspection; conservation; and animal and plant health programs. The main mandatory spending items are the Supplemental Nutrition Assistance Program (SNAP), child nutrition, crop insurance, and the farm commodity and conservation programs paid by the Commodity Credit Corporation.

* This is an edited, reformatted and augmented version of Congressional Research Service Publication No. R45230, Updated December 7, 2018.

For FY2019, both the House and Senate Appropriations Committees reported agriculture bills (H.R. 5961, S. 2976) in May 2018. The Senate amended and passed its version as Division C of a four-bill minibus on August 1, 2018 (H.R. 6147). In the absence of a final appropriation, Congress enacted two continuing resolutions through December 21, 2018 (P.L. 115-245, Division C; and H.J.Res 143).

The Trump Administration requested $17 billion for discretionary-funded accounts within the jurisdiction of Agriculture appropriations, a reduction of $6.2 billion from FY2018 (-27%). Both the House-reported and Senate-passed bills generally reject most of the proposed reductions.

The discretionary total of the House-reported bill is $23.23 billion, which would be $14 million less than enacted in FY2018 (-0.1%). The discretionary total of the Senate-passed bill is also $23.23 billion. However, the Senate bill's total would be $229 million more than enacted in FY2018 (+1%) on a comparable basis that excludes the CFTC. The Senate-passed bill would provide about $250 million more than the House-passed bill on a comparable basis.

The primary changes at the agency level that comprise the differences between the bills and from FY2018 include the following: Both the House and Senate bills would increase FDA appropriations (+$308 million in the House bill; +$159 million in the Senate bill). Both bills increase appropriations for animal and plant health programs (+$16 million to +$19 million). The House bill would provide more base funding for rural water and waste disposal (+$81 million), but none of the extra money that was provided separately in FY2018. The Senate bill would not change the base funding for rural water but continues $400 million of the extra funding from last year. For rural broadband, both the House and Senate bills would continue extra funding from a FY2018 pilot ($550 million in the House bill; $425 million in the Senate bill). The House bill would increase appropriations for agricultural research (+$79 million), and the Senate bill would increase Agricultural Research Service programming (+98 million) but would not provide any money for construction (-$141 million). Both bills provide less for WIC (-$175 million in the House bill, and -$25 million in the Senate bill), though the Senate bill has a larger rescission from prior-year WIC funds than does the House bill. The House bill would reduce base funding for the international Food for Peace program (-$100 million) and does not renew extra funding provided last year (-$116 million), while the Senate bill would keep it constant overall.

The appropriations also carry mandatory spending—largely determined in separate authorizing laws—that would total $122 billion. Thus, the overall total of the both bills is about $145 billion.

Both bills contain policy provisions affecting disaster programs, rural definitions, industrial hemp, animal products, nutrition programs, dietary guidelines, CFTC, and tobacco products.

STATUS OF FY2019 AGRICULTURE APPROPRIATIONS

Both the House and Senate Appropriations Committees have reported Agriculture appropriations bills for FY2019 (H.R. 5961, S. 2976). The Senate amended and passed its version as Division C of a four-bill minibus (H.R. 6147). In the absence of a final appropriation, FY2019 began under two continuing resolutions that last until December 21 (*Table 1*).

Specifically, the House Appropriations Committee passed its bill, H.R. 5961, on May 16, 2018, after the Agriculture Subcommittee passed its markup on May 9. The Senate Appropriations Committee passed its bill, S. 2976, on May 24, 2018, after the Agriculture Subcommittee passed its markup on May 22. The full Senate passed a four-bill minibus containing agriculture appropriations on August 1, 2018 (*Figure 1, Appendix B*).

Higher discretionary budget caps in the Bipartisan Budget Act of 2018 (P.L. 115-123) may facilitate enactment of a final appropriation. The discretionary total of the House-reported bill is $23.23 billion and would be $14 million less than enacted in FY2018 (-0.1%; *Table 2*). The discretionary total of the Senate-passed bill is also $23.23 billion. However, the Senate bill's total would be $229 million more than enacted in FY2018 (+1%) on a comparable basis that excludes the Commodity Futures Trading Commission (CFTC).[1] The Senate-passed bill would provide about $250 million more than the House-passed bill on a comparable basis.

The appropriations also carry mandatory spending—though that is largely determined in separate authorizing laws—that would total nearly $122 billion. Thus, the overall total of each of the proposed bills is about $145 billion.

---

[1] Jurisdiction for CFTC appropriations differs between the chambers. Since FY2008, CFTC is marked up in the Agriculture Subcommittee of the House Appropriations Committee and in the Financial Services and General Government Subcommittee of the Senate Appropriations Committee. The enacted CFTC appropriation is carried in the Agriculture bill in even-numbered fiscal years and in the Financial Services bill in odd-numbered fiscal years.

Table 1. Status of FY2019 Agriculture Appropriations

House Action			Senate Action			Continuing Resolution	Final Appropriation
Subcmte.	Cmte.	Floor	Appropriation	Cmte.	Floor		
5/9/2018 Drafta Voice vote	5/16/2018 H.R. 5961 H.Rept. 115-706 Vote 31-20	—	5/22/2018 Draft Voice vote	5/24/2018 S. 2976 S.Rept. 115-259 Vote 31-0	8/1/2018 H.R. 6147 Division C Vote 92-6	12/7/2018 H.J.Res 143 until 12/21/18 P.L. 115-245, Division C)	—

Source: CRS, compiled from the Legislative Information System.

[a] The House subcommittee draft is at https://docs.house.gov/meetings/AP/AP01/20180509/108287/BILLS115HR-SC-AP-FY2019-Agriculture-SubcommitteeDraft.pdf, and the committee report draft is athttps://docs.house. gov/meetings/AP/AP00/20180516/108312/HRPT-115-HR-FY2019-Agriculture.pdf.

Table 2. Agriculture and Related Agencies Appropriations, by Title, FY20 1 8-FY2019 (budget authority in millions of dollars)

	FY2018	FY2019				Change FY2018 to FY2019...	
Title of Agriculture Appropriations Act	P.L. 115-141	Admin. Request	House H.R. 5961	Senate H.R. 6147	Enac-ted	House	Senate
I. Agricultural Programs: Discretionary	5,622.8	4,752.9	5,704.5	5,587.8	—	+81.7	-35.1
Mandatory	*1,344.0*	*1,374.0*	*1,374.0*	*1,374.0*	—	*+30.0*	*+30.0*
Subtotal	*6,966.8*	*6,126.9*	*7,078.5*	*6,961.8*	—	*+111.7*	*-5.1*
II. Farm Production and Conservation Programs	2,735.6	2,197.9	2,760.3	2,728.0	—	+24.7	-7.6
Mandatory	*23,198.3*	*24,097.5*	*24,097.5*	*24,097.5*	—	*+899.2*	*+899.2*
Subtotal	*25,933.9*	*26,295.4*	*26,857.8*	*26,825.5*	—	*+923.9*	*+891.6*
III. Rural Development	3,000.9[a]	2,022.5	3,078.5[a]	3,000.9[a]	—	+77.6	+0.0
IV. Domestic Food Programs: Discretionary	6,709.8	5,990.1	6,528.5	6,696.1	—	-181.2	-13.7
Mandatory	*98,209.6*	*96,342.3*	*96,344.8*	*96,344.8*	—	*-1,864.9*	*-1,864.9*
Subtotal	*104,919.4*	*102,332.3*	*102,873.3*	*103,040.9*	—	*-2,046.1*	*-1,878.5*
V. Foreign	2,021.0[a]	205.0	1,925.8	2,152.3	—	-95.2	+131.4

	FY2018	FY2019				Change FY2018 to FY2019...	
Title of Agriculture Appropriations Act	P.L. 115-141	Admin. Request	House H.R. 5961	Senate H.R. 6147	Enacted	House	Senate
Assistance VI. Related Agencies: Food and Drug Administration	2,811.9[a]	3,183.7	3,119.6	2,970.9	—	+307.7	+159.0
Commodity Futures Trading Commission	249.0	281.5	255.0	[281.5][b]	—	+6.0	[+32.5][b]
VII. General Provisions:							
CHIMPS and rescissions[c]	-801.0	-1,197.8	-315.0	-395.0	—	+486.0	+406.0
Other appropriations	1,378.1	0.0	579.5	889.5	—	-798.6	-488.6
Scorekeeping adjustments[d]	-481.0	-409.0	-404.0	-404.0	—	+77.0	+77.0
Discretionary: Senate basis w/o CFTC	[22,998.0]	16,745.2	[22,977.8]	23,226.5	—	-20.2	+228.5
Discretionary: House basis w/ CFTC	23,247.0	17,026.7	23,232.8	[23,508.0][b]	—	-14.2	+261.0[b]
Mandatory	*122,752.0*	*121,813.8*	*121,816.3*	*121,816.3*	—	-935.7	-935.7
Total: Senate basis w/o CFTC	*[145,750.0]*	*138,559.0*	*[144,794.0]*	*145,042.8*	—	-955.9	-707.2
Total: House basis w/ CFTC	*145,999.0*	*138,840.5*	*145,049.0*	[145,324.3][b]	—	-949.9	-674.7[b]

Source: CRS, using appropriations text and reports and unpublished CBO tables.

Notes: Amounts are nominal discretionary budget authority in millions of dollars unless labeled otherwise. Excludes amounts in supplemental appropriations acts. Bracketed amounts are not in the official totals due to differing House-Senate jurisdiction for CFTC.

[a] Excludes a portion of the other appropriations that are provided separately in General Provisions.

[b] See Division B of H.R. 6147.

[c] Changes in Mandatory Program Spending (CHIMPS) are reductions made to mandatory programs via appropriations. Rescissions are permanent cancellations of previously provided budget authority.

[d] "Scorekeeping adjustments" are not necessarily appropriated but are part of the official CBO accounting.

The Trump Administration released its FY2019 budget request on February 12, 2018,[2] along with the detailed justification from the U.S. Department of Agriculture (USDA).[3] The Administration requested $17 billion of discretionary Agriculture appropriations (including CFTC), which would have been a reduction of $6.2 billion from FY2018 (-27%).

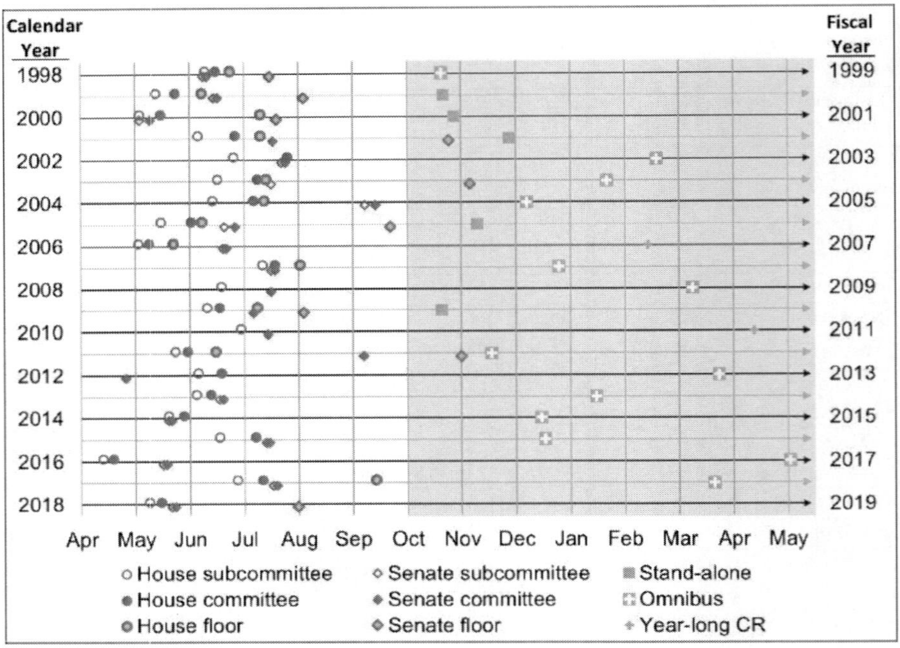

Source: CRS.

Figure 1. Timeline of Action on Agriculture Appropriations, FY1999-FY2019.

Scope of Agriculture Appropriations

The Agriculture appropriations bill—formally known as the Agriculture, Rural Development, Food and Drug Administration, and

[2] Office of Management and Budget (OMB), *FY2019 Budget of the U.S. Government*, especially in the Appendix, https://www.whitehouse.gov/omb/appendix.

[3] USDA, *FY2019 USDA Budget Summary*; and USDA, *2019 Explanatory Notes*, http://www.obpa.usda.gov.

Related Agencies Appropriations Act—funds all of USDA, excluding the U.S. Forest Service. It also funds the Food and Drug Administration (FDA) in the Department of Health and Human Services. In even-numbered fiscal years, the act carries CFTC funding under a practice started in FY2008 for handling House-Senate jurisdictional differences.

Jurisdiction is with the House and Senate Committees on Appropriations and their respective Subcommittees on Agriculture, Rural Development, Food and Drug Administration, and Related Agencies. The bill includes mandatory and discretionary spending, but the discretionary amounts are the primary focus during the bill's development. The scope of the bill is shown in *Figure 2.*

The federal budget process treats discretionary and mandatory spending differently:[4]

- Discretionary spending is controlled by annual appropriations acts and receives most of the attention during the appropriations process. The annual budget resolution[5] process sets spending limits for discretionary appropriations. Agency operations (salaries and expenses) and many grant programs are discretionary.
- Mandatory spending[6]—though carried in the appropriation and usually advanced unchanged—is controlled by budget rules during the authorization process.[7] Spending for so-called entitlement programs is set in laws such as the 2014 farm bill[8] and 2010 child nutrition reauthorizations.[9]

In the FY2018 Agriculture appropriations, discretionary appropriations were 16% ($23.3 billion) of the $146 billion total. Mandatory spending

---

[4] See CRS Report R44582, *Overview of Funding Mechanisms in the Federal Budget Process, and Selected Examples.*

[5] See CRS Report R42388, *The Congressional Appropriations Process: An Introduction.*

[6] Mandatory spending creates funding stability and consistency compared to annual discretionary appropriations. In agriculture, it originated with farm commodity programs that had uncertain outlays due to weather and markets.

[7] See CRS Report 98-560, *Baselines and Scorekeeping in the Federal Budget Process.*

[8] P.L. 113-79. See CRS In Focus IF10783, *Farm Bill Primer: Budget Issues.*

[9] P.L. 111-296. See CRS Report R44373, *Tracking the Next Child Nutrition Reauthorization: An Overview.*

carried in the act comprised $123 billion, about 84% of the total. About $99 billion of the $123 billion mandatory amount could be attributed to programs in the 2014 farm bill (*Figure 2*). Some programs are not in the authorizing jurisdiction of the House or Senate Agriculture Committees.

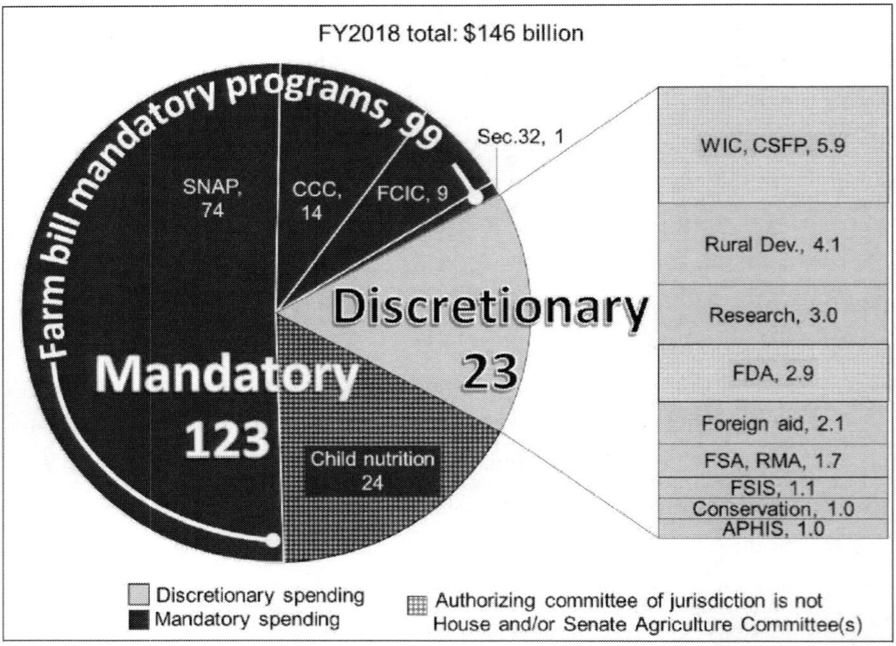

Source: CRS. Does not show some agencies under $0.5 billion. Includes General Provisions with agencies.

Notes: SNAP = Supplemental Nutrition Assistance Program; CCC = Commodity Credit Corporation; FCIC = Federal Crop Insurance Corporation; Section 32 = Funds for Strengthening Markets, Income and Supply; WIC = Special Supplemental Nutrition Program for Women, Infants, and Children; CSFP = Commodity Supplemental Food Program; FDA = Food and Drug Administration; FSA = Farm Service Agency; RMA = Risk Management Agency; FSIS = Food Safety and Inspection Service; APHIS = Animal and Plant Health Inspection Service.

Figure 2. Scope of Agriculture and Related Agencies Appropriations. (FY2018 budget authority in billions of dollars)

Key Budget Terms

Budget authority is the main purpose of an appropriations act or a law authorizing mandatory spending. It provides the legal basis from which to obligate funds. It expires at the end of a period and is usually available for one year unless specified otherwise (e.g., two years or indefinite). Most amounts in this report are budget authority.

Obligations are contractual agreements between an agency and its clients or employees. They occur when an agency proceeds to spend money from its budget authority. The Antideficiency Act prohibits agencies from obligating more budget authority than is provided in law, such as during a government shutdown.

Outlays are the payments (cash disbursements) that satisfy a valid obligation. Outlays may differ from budget authority or obligations because payments from an agency may not occur until services are fulfilled, goods are delivered, or construction is completed, even though an obligation occurred.

Program level represents the sum of the activities supported or undertaken by an agency. A program level may be higher than a budget authority if the program (1) receives *user fees* that can be used to pay for activities, (2) makes or guarantees *loans* that are leveraged on the expectation of repayment (more than $1 of loan authority for $1 of budget authority), or (3) receives *transfers* from other agencies.

Rescissions are adjustments that cancel or reduce budget authority after enactment. They score budgetary savings.

CHIMPS (Changes in Mandatory Program Spending) are adjustments in an appropriations act to mandatory budget authority. CHIMPS in appropriations usually reduce or limit spending by mandatory programs for one year and score budgetary savings. They do not change the underlying authority of the program in statute.

Within the discretionary total, the largest discretionary spending items are the Special Supplemental Nutrition Program for Women, Infants, and Children (WIC); rural development; agricultural research; FDA; foreign food aid and trade; farm assistance program salaries and loans; food safety inspection; conservation; and animal and plant health programs (*Figure* 2).

The main mandatory spending items are the Supplemental Nutrition Assistance Program (SNAP, and other food and nutrition act programs), child nutrition (school lunch and related programs), crop insurance, and farm commodity and conservation programs that are paid through USDA's Commodity Credit Corporation (CCC). SNAP is referred to as an "appropriated entitlement" and requires an annual appropriation.[10] Amounts for the nutrition program are based on projected spending needs. In contrast, the CCC operates on a line of credit. The annual appropriation provides funding to reimburse the Treasury for the use of this line of credit.

RECENT TRENDS IN AGRICULTURE APPROPRIATIONS

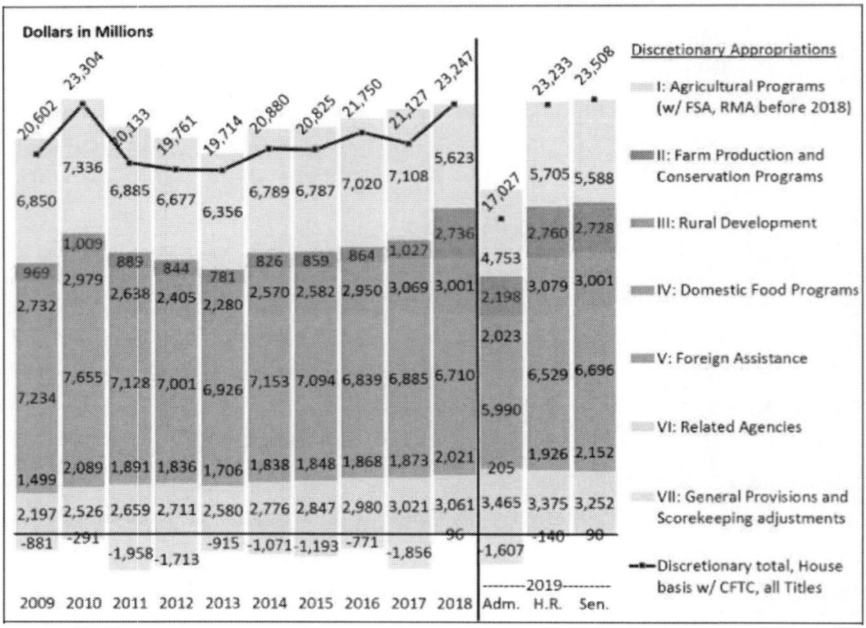

Source: CRS.

Note: For FY2019, Adm. is the Administration's request, H.R. is committee-reported H.R. 5961 and Sen. is Senate-passed H.R. 6147. Includes CFTC in Related Agencies in all years.

Figure 3. Discretionary Agriculture Appropriations, by Title, FY2009-FY2019.

---

[10] See CRS Report RS20129, *Entitlements and Appropriated Entitlements in the Federal Budget Process.*

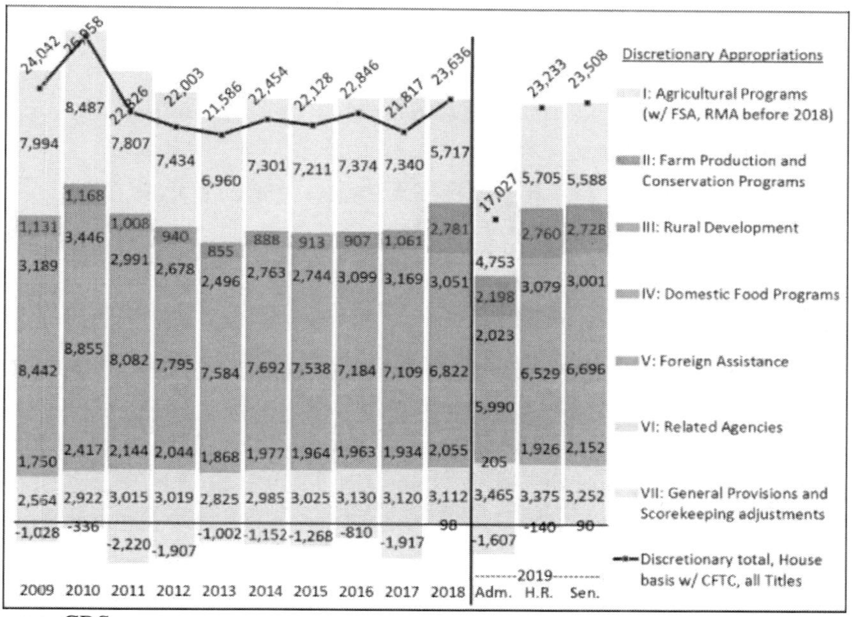

Source: CRS.

Note: For FY2019, Adm. is the Administration's request, H.R. is committee-reported H.R. 5961 and Sen. is Senate-passed H.R. 6147. Includes CFTC in Related Agencies in all years. Budget authority is millions of dollars adjusted to FY2019 using the gross domestic product price deflator.

Figure 4. Inflation-Adjusted Agriculture Appropriations, FY2009-FY2019.

Over time, changes by title of the Agriculture appropriations bill have generally been proportionate to changes in the bill's total discretionary limit, though some activities have sustained relative increases and decreases. Agriculture appropriations peaked in FY2010, declined through FY2013, and since then have increased (*Figure 3*).

Comparisons to historical benchmarks, though, may be affected by adjustments for inflation (*Figure 4*). In FY2018, USDA reorganization affected the placement of some programs between Titles I and II of the bill.

The stacked bars in *Figure 3* represent the discretionary authorization for each appropriations title. The total of the positive stacked bars is the budget authority in Titles I-VI. Prior to FY2018, it was higher than the official discretionary spending allocation (the line) because of the budgetary offset from negative amounts in Title VII (general provisions) and other scorekeeping adjustments that were negative mostly due to limits on mandatory programs and rescissions.

ACTION ON FY2019 APPROPRIATIONS

Administration's Budget Request

The Trump Administration released its FY2019 budget request on February 12, 2018,[11] one week later than the usual first Monday in February. USDA concurrently released its more detailed budget summary and justification,[12] as did the FDA,[13] and the independent agencies of the CFTC[14] and the Farm Credit Administration (FCA).[15] The Administration also highlighted some of the proposed reductions and eliminations separately.[16]

From these documents, the congressional appropriations committees evaluated the request and began to consider their own bills in the spring of 2018. For accounts in the jurisdiction of the Agriculture appropriations bill, the Administration's budget proposed $17.4 billion, a 25% reduction from FY2018 (*Table 2, Figure 3*).

The timing of the Administration's budget request for FY2019 preceded Congress enacting the final, omnibus FY2018 appropriation in March 2018. Therefore, amounts in the FY2018 column of the Administration's budget documents are based on FY2017 and the continuing resolution and are not reliable indicators of the enacted FY2018 levels that came later.

---

[11] OMB, *FY2019 Budget of the U.S. Government*, Appendix, https://www.whitehouse.gov/omb/appendix.

[12] USDA, *FY2019 USDA Budget Summary*; and USDA, *2019 Budget Explanatory Notes*, http://www.obpa.usda.gov.

[13] FDA, *FY2019 FDA Justification of Estimates for Appropriations*, https://www.hhs.gov/about/budget/index.html.

[14] CFTC, *FY2019 CFTC Budget Request*, https://www.cftc.gov/sites/default/files/idc/groups/public/documents/file/cftcbudget2019.pdf.

[15] FCA, *FY2019 FCA Proposed Budget*, https://www.fca.gov/Download/BudgetFY2019.pdf.

[16] OMB, *FY2019 Budget of the U.S. Government, Major Savings and Reforms.*

Discretionary Budget Caps and Subcommittee Allocations

Budget enforcement for appropriations has both procedural and statutory elements.

The procedural elements are associated with the budget resolution and are enforced through points of order. Typically, each chamber's full Appropriations Committee receives a top-line procedural limit on discretionary budget authority, referred to as a "302(a)" allocation, from the Budget Committee via an annual budget resolution passed by each chamber. The Appropriations Committees then each in turn subdivide their allocation among their subcommittees, referred to as the "302(b)" allocations.[17]

The statutory elements impose limits on discretionary spending in FY2012-FY2021 and are enforced through discretionary budget caps and sequestration (2 U.S.C. 901(c)).[18] The Budget Control Act of 2011 (BCA, P.L. 112-25) set discretionary budget caps through FY2021 as a way of reducing federal spending.[19] Bipartisan Budget Acts (BBAs) in 2013, 2015, and 2018 (P.L. 113-67, P.L. 114-74, and P.L. 115-123, respectively) have avoided sequestration on discretionary spending—with the exception of FY2013—by raising those caps (*Appendix A*).[20]

In February 2018, the BBA raised the FY2019 cap on nondefense discretionary spending by $68 billion and the cap on defense spending by $85 billion.[21] It also provided language to execute (or "deem") those higher caps for the appropriations process without following the usual procedures for an FY2019 budget resolution.[22]

---

[17] CRS Report R42388, *The Congressional Appropriations Process: An Introduction.*

[18] CRS Report R42972, *Sequestration as a Budget Enforcement Process: Frequently Asked Questions.*

[19] CRS Report R43411, *The Budget Control Act of 2011: Legislative Changes to the Law and Their Budgetary Effects.*

[20] Looking to the future, though, the budget caps for FY2020 and FY2021 remain unchanged from the original BCA in 2011. They are less than the revised caps for FY2019. Unless the caps change, this implies less future spending.

[21] CRS Insight IN10861, *Discretionary Spending Levels Under the Bipartisan Budget Act of 2018.*

[22] CRS Report R45157, *The Bipartisan Budget Act of 2018 and an FY2019 Budget Resolution.*

Under these higher caps and authorities, the House and Senate Appropriations Committees proceeded to mark up the FY2019 appropriations bills. The Agriculture appropriations bills are receiving roughly the same subcommittee allocation ("302(b)" allocation) in each chamber that was used to complete the FY2018 appropriation under the BBA. For FY2019, the subcommittee allocations for agriculture appropriations are

- House: $23.273 billion (H.Rept. 115-710, May 23, 2018), including the CFTC.
- Senate: $23.235 billion (S.Rept. 115-260, May 24, 2018), excluding the CFTC.

House Action

The House Agriculture Appropriations Subcommittee marked up its FY2019 bill on May 9, 2018, by voice vote.[23] On May 16, 2018, the full Appropriations Committee passed and reported an amended bill (H.R. 5961, H.Rept. 115-706) by a vote of 31-20 (*Table 1, Figure 1*).

The $23.23 billion discretionary total in the House-reported FY2019 Agriculture appropriation would be $14 million less than enacted in FY2018 (-0.1%; *Table 2, Figure 3*). Generally speaking, the House-reported bill does not include most of the reductions proposed by the Administration and continues the trend of appropriations from prior years.

Table 3 provides details at the agency level. The primary changes from FY2018 at the agency level that comprise the relatively flat $14 million overall decrease in the House-reported bill would do the following:

---

[23] The House subcommittee draft is at https://docs.house.gov/meetings/AP/AP01/20180509/108287/BILLS-115HRSC-AP-FY2019-Agriculture-SubcommitteeDraft.pdf, and the draft of the committee report is at https://docs.house.gov/meetings/AP/AP00/20180516/108312/HRPT-115-HR-FY2019-Agriculture.pdf.

- Increase base FDA appropriations by $308 million (+11%). However, it does not continue to separate FDA funding for the opioid crisis that was in the General Provisions title of the FY2018 appropriation (-$94 million).
- Increase agricultural research (+$79 million, +2.6%) by raising appropriations for the Agricultural Research Service and National Institute of Agriculture.
- Increase the Animal and Plant Health Inspection Service (APHIS) by $16 million (+1.7%).
- Increase the Rural Utility Service by $82 million (+13%), mostly in the base funding for rural water and waste disposal programs. However, it does not continue separate funding for rural water that was in the General Provisions title last year (-$500 million) or for telemedicine for the opioid crisis (-$20 million). It reduces separate funding for a broadband pilot program while continuing to fund some of it through the General Provisions title (-$50 million).
- Decrease discretionary appropriations for domestic nutrition assistance programs by reducing WIC by $175 million (-2.8%) and the commodity assistance programs by $15 million (-4.7%). However, the reduction scored in the General Provisions title by rescinding WIC carryover balances is smaller in FY2019, retaining more budget authority (+$500 million).
- Decrease the base funding in international food assistance Food for Peace grants by $100 million (-6.2%) and not continue the extra funding that was in the General Provisions title of the FY2018 appropriations (-$116 million).

In addition to discretionary spending, the appropriations also carry funding for mandatory spending—largely determined in separate authorizing laws—that would total $121.82 billion, about $936 million less than in FY2018 because of automatic changes in economic conditions and entitlement enrollment rather than any change from congressional action. Thus, the overall total of the House-reported bill is about $145 billion.

Senate Action

The Senate Agriculture Appropriations Subcommittee marked up a FY2019 bill on May 22, 2018, by voice vote. On May 24, 2018, the Appropriations Committee passed and reported an amended bill (S. 2976, S.Rept. 115-259) by a vote of 31-0. On August 1, 2018, the Senate passed a four-bill minibus (H.R. 6147) by a vote of 92-6, with agriculture as Division C (*Table 1, Figure 1*).

The discretionary total of the Senate-passed bill is also $23.23 billion. However, the Senate bill's total would be $229 million more than enacted in FY2018 (+1%) on a comparable basis that excludes the CFTC, since the latter was part of the enacted FY2018 appropriation. The Senate-passed bill would provide about $250 million more than the House-reported bill would on a comparable basis computed by subtracting CFTC from the House bill.

Table 3 provides details at the agency level. The primary changes from FY2018 at the agency level that comprise the Senate-passed bill's overall $228 million increase are the following:

- Increase base FDA appropriations by $159 million (+6%). Like the House bill, it does not continue separate FDA funding for the opioid crisis that was in the General Provisions title of the FY2018 appropriation (-$94 million).
- Increase APHIS by $19 million (+1.9%), slightly more than the House bill.
- Decrease discretionary appropriations for domestic nutrition assistance programs by reducing WIC by $25 million (-0.4%). This is a smaller reduction than in the House bill. In the rescissions in the General Provisions title, the reduction is smaller than was rescinded in FY2019 (+$400 million). Nonetheless, the Senate bill's rescission is greater than in the House bill.
- Increase international nutrition assistance by raising the base funding for Food for Peace grants by $116 million (+7.2%). The

extra funding that was in the General Provisions title of the FY2018 appropriations is not continued (-$116 million).

- Decrease the four agricultural research agencies by $44 million (-1.4%), mostly by providing no funding for ARS building and facilities (-$141 million), while increasing ARS salaries and expenses (+$98 million).
- Decrease the extra funding for rural development compared to the amount provided in the FY2018 General Provisions (-$100 million for rural water, -$175 million for broadband). Base funding for rural development is unchanged overall.

The Senate-passed bill's mandatory spending is virtually identical to that of the House-reported bill ($121.82 billion). Its overall total of discretionary and mandatory appropriations is $145 billion.

Continuing Resolutions

In the absence of a final FY2019 Agriculture appropriation, Congress enacted two continuing resolutions (CRs) to continue government operations beyond October 1, 2018—P.L. 115-245, Division C (through 12/7/2018), and H.J.Res 143 (through 12/21/2018).[24]

In general, a CR continues the funding rates and conditions that were in the previous year's appropriation.[25] The Office of Management and Budget (OMB) may prorate funding to the agencies on an annualized basis for the duration of the CR through a process known as apportionment.[26] For the 81 days (22%) of FY2019 through December 21, the CRs

- continue the terms of the FY2018 Agriculture appropriations act (Section 101) and exclude the FY2018 change in mandatory

---

[24] CRS Report R42647, *Continuing Resolutions: Overview of Components and Recent Practices.*

[25] CRS Report RL34700, *Interim Continuing Resolutions (CRs): Potential Impacts on Agency Operations.*

[26] For example, if a CR lasts for three months, then 3/12 of the previous fiscal year amount may be apportioned to limit agency spending. See OMB, "Apportionment of the Continuing Resolution(s) for Fiscal Year 2019," September 28, 2018, https://www.whitehouse.gov/wp-content/uploads/2017/11/bulletin_no_18_05.pdf.

program spending (CHIMP) on the Biomass Crop Assistance Program, which is not authorized for FY2019; and

- provide sufficient funding to maintain mandatory program levels, including for nutrition programs (Section 110)—this is standard language for recent years' CRs but is additionally important for SNAP, because some authorizations in the 2014 farm bill began expiring after FY2018.[27]

CRs may adjust prior-year amounts through anomalies or make specific administrative changes. Five anomalies specifically apply to the agriculture appropriation in both CRs:

1. *Child Nutrition Programs:* apportionment for a summer foods program sufficient to allow it to be operational by May 2019 (Section 114).
2. *Rural Utilities Service:* allows a loan authorization level for the Rural Water and Waste Disposal program of $4.141 billion (Section 115).
3. *Commodity Credit Corporation (CCC):* allows CCC to receive its transfer about a month earlier than usual, prior to a final report and audit. Many farm bill payments to farmers are due in October 2018 in addition to USDA's plan to make supplemental payments under a trade assistance program.[28] Without the anomaly, CCC might exhaust its $30 billion line of credit (Section 116).
4. *Agricultural Research Service:* an additional $42 million for operations and maintenance at the National Bio and Agro-Defense Facility (NBAF) being built in Manhattan, Kansas, and being transferred to USDA from the Department of Homeland Security (Section 117).
5. *Department of Homeland Security (DHS):* DHS may transfer up to $15 million to USDA to support NBAF operations (Section 125).

---

[27] CRS In Focus IF10989, *Expiration of the 2014 Farm Bill: Some Potential Implications.*

[28] CRS Report R45310, *Farm Policy: USDA's Trade Aid Package.*

Table 3. Agriculture and Related Agencies Appropriations, by Agency, FY2016-FY2019 (budget authority in millions of dollars)

Agency or Major Program	FY2016	FY2017	FY2018		FY2019		Change FY18 to FY19		
	P.L. 114-113	P.L. 115-31	P.L. 115-141	Admin. Request	House H.R. 5961	Senate H.R. 6147	Enacted	House	Senate
Title I. Agricultural Programs									
Departmental Administration	373.2	403.9	396.0	369.8	387.8	389.5	—	-8.1	-6.5
Research, Education and Economics									
Agricultural Research Service	1,355.9	1,269.8	1,343.4	1,019.0	1,395.9	1,301.0	—	+52.6	-42.4
National Institute of Food and Agriculture	1,326.5	1,362.9	1,407.8	1,257.7	1,452.6	1,423.2	—	+44.8	+15.4
National Agricultural Statistics Service	168.4	171.2	191.7	165.0	173.7	174.8	—	-18.0	-17.0
Economic Research Service	85.4	86.8	86.8	45.0	86.8	86.8	—	+0.0	+0.0
Under Secretary	0.9	0.9	0.8	0.8	0.8	0.8	—	+0.0	+0.0
Marketing and Regulatory Programs									
Animal and Plant Health Inspection Service	897.6	949.4	985.1	742.0	1,001.5	1,003.7	—	+16.5	+18.6
Agricultural Marketing Service[a]	82.5	86.2	152.8	119.7	154.3	157.1	—	+1.5	+4.3
Section 32 (M)	1,303.0	1,322.0	1,344.0	1,374.0	1,374.0	1,374.0	—	+30.0	+30.0
Grain Inspection, Packers, Stockyards Admin.[a]	43.1	43.5	moved into Agricultural Marketing Servicea						
Under Secretary	0.9	0.9	0.9	0.8	0.9	0.9	—	+0.0	+0.0
Food Safety									
Food Safety and Inspection Service	1,014.9	1,032.1	1,056.8	1,032.3	1,049.3	1,049.3	—	-7.5	-7.5
Under Secretary	0.8	0.8	0.8	0.8	0.8	0.8	—	+0.0	+0.0

Table 3. (Continued)

Agency or Major Program	FY2016	FY2017	FY2018		FY2019		Change FY18 to FY19		
	P.L. 114-113	P.L. 115-31	P.L. 115-141	Admin. Request	House H.R. 5961	Senate H.R. 6147	Enacted	House	Senate
Farm and Commodity Programs[a]									
Farm Service Agency[b]	1,595.1	1,624.0	moved to Title II: Farm Production and Conservation[a]						
*FSA Farm Loans: Loan Authority*c	*6,402.1*	*8,002.6*	moved to Title II: Farm Production and Conservationa						
Risk Management Agency Salaries and Exp.	74.8	74.8	moved to Title II: Farm Production and Conservation[a]						
Federal Crop Insurance Corporation (M)	7,858.0	8,667.0	moved to Title II: Farm Production and Conservation[a]						
Commodity Credit Corporation (M)	6,871.1	21,290.7	moved to Title II: Farm Production and Conservationa						
Under Secretary	0.9	0.9	moved to Title II: Farm Production and Conservationa						
Subtotal									
Discretionary	7,020.3	7,107.7	5,622.8	4,752.9	5,704.5	5,593.8	—	+81.7	-29.1
Mandatory (M)	*16,032.6*	*31,280.2*	*1,344.0*	*1,374.0*	*1,374.0*	*1,374.0*	—	*+30.0*	*+30.0*
Subtotal	*23,052.9*	*38,387.9*	*6,966.8*	*6,126.9*	*7,078.5*	*6,967.8*	—	*+111.7*	*+0.9*
Title II. Farm Production and Conservation[a]									
Business Center	—	—	1.0	196.4	115.4	1.0	—	+114.4	+0.0
Farm Service Agency[b]	moved from Title I[a]		1,625.2	1,294.2	1,518.8	1,622.7	—	-106.4	-2.6
*FSA Farm Loans: Loan Authority*c	moved from Title		*8,005.6*	*7,617.7*	*7,987.7*	*8,017.7*	—	*-17.9*	*+12.1*

Agency or Major Program	FY2016	FY2017	FY2018		FY2019		Change FY18 to FY19		
	P.L. 114-113	P.L. 115-31	P.L. 115-141	Admin. Request	House H.R. 5961	Senate H.R. 6147	Enacted	House	Senate
	I[a]								
Risk Management Agency Salaries and Exp.	moved from Title I[a]		74.8	37.9	75.4	74.8	—	+0.6	+0.0
Federal Crop Insurance Corporation (M)	moved from Title I[a]		8,913.0	8,687.0	8,687.0	8,687.0	—	-226.0	-226.0
Commodity Credit Corporation (M)	moved from Title I[a]		14,284.8	15,410.0	15,410.0	15,410.0	—	+1,125.2	+1,125.2
Conservation Operations	850.9	864.5	874.1	669.0	890.3	879.1	—	+16.2	+5.0
Watershed and Flood Prevention	—	150.0	150.0	—	150.0	150.0	—	+0.0	+0.0
Watershed Rehabilitation Program	12.0	12.0	10.0	—	10.0	—	—	+0.0	-10.0
Under Secretary	0.9	0.9	0.9	0.9	0.9	0.9	—	+0.0	+0.0
Subtotal									
Discretionary	863.8	1,027.4	2,735.6	2,197.9	2,760.3	2,728.0	—	+24.7	-7.6
Mandatory (M)	moved from Title I I[a]		*23,198.3*	*24,097.5*	*24,097.5*	*24,097.5*	—	*+899.2*	*+899.2*
Subtotal	moved from Title I I[a]		*25,933.9*	*26,295.4*	*26,857.8*	*26,825.5*	—	*+923.9*	*+891.6*
Title III. Rural Development									
Salaries and Expenses (including transfers)[d]	682.9	675.8	680.8	585.9	686.8	682.8	—	+6.0	+2.0
Rural Housing Service	1,616.4	1,654.9	1,582.4	1,351.4	1,582.5	1,585.2	—	+0.2	+2.8
RHS Loan Authority[c]	*27,496.8*	*28,083.4*	*28,390.1*	*27,760.0*	*28,345.5*	*28,590.1*	—	*-44.6*	*+200.0*
Rural Business-Cooperative Service[e]	90.5	97.7	109.5	0.0	99.5	104.2	—	-10.0	-5.4
RBCS Loan Authority[c]	*979.3*	*988.4*	*991.2*	*0.0*	*1,027.5*	*991.2*	—	*+36.3*	*+0.0*

Table 3. (Continued)

Agency or Major Program	FY2016	FY2017	FY2018		FY2019		Change FY18 to FY19		
	P.L. 114-113	P.L. 115-31	P.L. 115-141	Admin. Request	House H.R. 5961	Senate H.R. 6147	Enacted	House	Senate
Rural Utilities Service	559.3	639.9	628.1[f]	85.2	709.6[f]	628.7[f]	—	+81.5	+0.6
RUS Loan Authority[c]	*8,210.6*	*8,217.0*	*8,219.9*	*7,413.1*	*8,419.9*	*8,219.9*	—	*+200.0*	*+0.0*
Under Secretary	0.9	0.9	moved to Departmental Administration as an Assistant to the Secretary[a]						
Subtotal, Discretionary	2,950.0	3,069.2	3,000.9[f]	2,022.5	3,078.5[f]	3,000.9[f]	—	+77.6	+0.0
Subtotal, RD Loan Authority[c]	*36,686.7*	*37,288.9*	*37,601.2*	*35,173.1*	*37,792.9*	*37,801.2*	—	+191.6	+200.0
Title IV. Domestic Food Programs									
Child Nutrition Programs (M)	22,149.7	22,794.0	24,254.1	23,146.9	23,183.5	23,184.0	—	-1,070.6	-1,070.1
WIC Program	6,350.0	6,350.0	6,175.0	5,750.0	6,000.0	6,150.0	—	-175.0	-25.0
SNAP, Food and Nutrition Act Programs (M)	80,849.4	78,480.7	74,013.5	73,218.3	73,219.3	73,219.3	—	-794.2	-794.2
Commodity Assistance Programs	296.2	315.1	322.1	55.5	306.9	322.1	—	-15.2	+0.0
Nutrition Programs Administration	150.8	170.7	153.8	160.8	162.8	164.7	—	+9.0	+10.8
Under Secretary	0.8	0.8	0.8	0.8	0.8	0.8	—	+0.0	+0.0
Subtotal									
Discretionary	6,838.9	6,884.7	6,709.8	5,990.1	6,528.5	6,696.1	—	-181.2	-13.7
Mandatory (M)	*102,958.1*	*101,226.7*	*98,209.6*	*96,342.3*	*96,344.8*	*96,344.8*	—	*-1,864.9*	*-1,864.9*
Subtotal	*109,797.0*	*108,111.3*	*104,919.4*	*102,332.3*	*102,873.3*	*103,040.9*	—	*-2,046.1*	*-1,878.5*
Title V. Foreign Assistance									
Foreign Agricultural Service	191.6	196.6	199.7	193.1	204.1	212.2	—	+4.5	+12.6
Food for Peace Title II, and admin. expenses	1,468.5[f]	1,466.1[f]	1,600.1[f]	0.1	1,500.1	1,716.1	—	-100.0	+116.0

Agency or Major Program	FY2016	FY2017	FY2018		FY2019		Change FY18 to FY19		
	P.L. 114-113	P.L. 115-31	P.L. 115-141	Admin. Request	House H.R. 5961	Senate H.R. 6147	Enacted	House	Senate
McGovern-Dole Food for Education	201.6	201.6	207.6	0.0	207.6	210.3	—	+0.0	+2.6
CCC Export Loan Salaries	6.7	8.5	8.8	7.1	9.2	8.8	—	+0.3	+0.0
Office of Codex Alimentarius	—	—	3.8	3.8	3.8	4.0	—	+0.0	+0.2
Under Secretary	—	—	0.9	0.9	0.9	0.9	—	+0.0	+0.0
Subtotal	1,868.5	1,872.9	2,021.0	205.0	1,925.8	2,152.3	—	-95.2	+131.4
Title VI. Related Agencies									
Food and Drug Administration	2,729.6	2,771.2	2,811.9	3,183.7	3,119.6	2,970.9	—	+307.7	+159.0
Commodity Futures Trading Commission[g]	250.0	[250.0]	249.0	281.5	255.0	[281.5][h]	—	+6.0	[+32.5]h
Subtotal	2,979.6	[3,021.2]	3,060.9	3,465.2	3,374.6	[3,252.4][h]	—	+313.7	[+191.5]h
Title VII. General Provisions									
Reductions in Mandatory Programs[i]									
a. Environmental Quality Incentives Program	-209.0	-179.0	—	-136.3	—	—	—	+0.0	+0.0
b. Watershed Rehabilitation Program	-68.0	-54.0	—	-46.2	—	—	—	+0.0	+0.0
c. Fresh Fruit and Vegetable Program	-125.0	-125.0	—	-125.0	—	—	—	+0.0	+0.0
d. Biorefinery Assistance Program	-19.0	-20.0	—	—	—	—	—	+0.0	+0.0
e. Biomass Crop Assistance Program	-20.0	-20.0	-21.0	—	—	—	—	+21.0	+21.0
f. Cushion of Credit (Rural Development)	-179.0	-132.0	—	-225.0	-50.0	—	—	-50.0	+0.0
g. Section 32	-216.0	-231.0	—	-342.0	—	—	—	+0.0	+0.0
h. Other CHIMPS and mandatory rescissions	+5.0	+17.0	+20.0	-36.4	+35.0	+5.0	—	+15.0	-15.0
Subtotal, CHIMPS	-831.0	-744.0	-1.0	-910.8	-15.0	+5.0	—	-14.0	+6.0
Rescissions (discretionary)	-34.0	-854.0	-800.0	-287.0	-300.0	-400.0	—	+500.0	+400.0
Other appropriations									

Table 3. (Continued)

Agency or Major Program	FY2016	FY2017	FY2018		FY2019		Change FY18 to FY19		
	P.L. 114-113	P.L. 115- 31	P.L. 115-141	Admin. Request	House H.R. 5961	Senate H.R. 6147	Enacted	House	Senate
a. Disaster/emergency programs	273.0	234.8[j]	—	—	—	—	—	+0.0	+0.0
b. Water and Waste Water	—	—	+500.0	—	—	+400.0	—	-500.0	-100.0
c. Broadband pilot	—	—	+600.0	—	+550.0	+425.0	—	-50.0	-175.0
d. Opioid Enforcement and Surveillance	—	—	+94.0	—	—	—	—	-94.0	-94.0
e. Food for Peace	250.0	134.0	+116.0	—	—	—	—	-116.0	-116.0
f. Other appropriations	33.1	103.4	68.1	—	+29.5	+64.5	—	-38.6	-3.6
Subtotal, Other appropriations	556.1	472.2	1,378.1	—	579.5	889.5	—	-798.6	-488.6
Total, General Provisions	-308.9	-1,125.8	577.1	-1,197.8	264.5	494.5	—	-312.6	-82.6
Scorekeeping Adjustments[k]									
Disaster declaration in this bill	-130.0	-206.1[j]	—	—	—	—	—	+0.0	+0.0
Other scorekeeping adjustments	-332.0	-524.0	-481.0	-409.0	-404.0	-404.0	—	+77.0	+77.0
Subtotal, Scorekeeping adjustments	-462.0	-730.1	-481.0	-409.0	-404.0	-404.0	—	+77.0	+77.0
Totals									
Discretionary: Senate basis w/o CFTC	[21,500.0]	20,877.0	[22,998.0]	16,745.2	[22,977.8]	23,226.5	—	-20.2	+228.5
Discretionary: House basis w/ CFTC	21,750.0	[21,127.0]	23,247.0	17,026.7	23,232.8	[23,508.0][h]	—	-14.2	[+261.0][h]
Mandatory (M)	*118,990.7*	*132,506.9*	*122,752.0*	121,813.8	121,816.3	121,816.3	—	-935.7	-935.7
Total: Senate basis w/o CFTC	*[140,490.7]*	*153,383.9*	*[145,750.0]*	138,559.0	[144,794.0]	145,042.8	—	-955.9	-707.2
Total: House basis w/ CFTC	*140,740.7*	*[153,633.9]*	*145,999.0*	138,840.5	145,049.0	[145,324][h]	—	-949.9	-674.7[h]

Source: CRS, using appropriations text and report tables, and unpublished CBO tables.

Notes: Amounts are nominal discretionary budget authority in millions of dollars unless labeled otherwise. (M) indicates that the account is mandatory authority (or primarily mandatory authority). Excludes amounts in supplemental appropriations acts. Bracketed amounts are not in the official totals due to differing House-Senate jurisdiction for CFTC but are shown for comparison.

[a] Row headings reflect recent USDA reorganization. The Farm Service Agency and Risk Management Agency were moved from Title I to Title II, as was the Commodity Credit Corporation and Federal Crop Insurance Corporation in mandatory spending. Grain Inspection, Packers, and Stockyards Administration was moved into the Agricultural Marketing Service.

[b] Includes regular FSA salaries and expenses, plus transfers for farm loan program salaries and administrative expenses. Also includes farm loan program loan subsidy, State Mediation Grants, Dairy Indemnity Program (mandatory funding), and Grassroots Source Water Protection Program. Does not include appropriations to the Foreign Agricultural Service for export loans and P.L. 480 administration that are transferred to FSA.

[c] Loan authority is the amount of loans that can be made or guaranteed with a loan subsidy. This amount is not added in the budget authority subtotals or totals.

[d] Includes Rural Development salaries and expenses and transfers from the three rural development agencies for salaries and expenses. Amounts for the agencies thus reflect program funds for loans and grants.

[e] Amounts for the Rural Business-Cooperative Service (RBCS) are before the rescission in the Cushion of Credit account, unlike in Appropriations Committee tables.

[f] The rescission is included with the CHIMPS as classified by CBO, which allows the RBCS subtotal to remain positive. f. Excludes a portion of the other appropriations that are provided separately in General Provisions.

[g] Jurisdiction for CFTC is in the House Agriculture Appropriations Subcommittee and the Senate Financial Services Appropriations Subcommittee. After FY2008, CFTC is carried in enacted Agriculture appropriations in even-numbered fiscal years in House Agriculture markup but not in Senate Agriculture markup. Bracketed amounts are not in the official totals due to differing House-Senate jurisdiction for CFTC but are shown for comparison.

[h] See Division B of H.R. 6147.

[I] Includes reductions (limitations and rescissions) to mandatory programs that may also be known as Changes in Mandatory Program Spending (CHIMPS).

[j] Includes $206 million appropriated for the Emergency Conservation Program (ECP) and Emergency Watershed Program (EWP) in the FY2017 second CR (P.L. 114-254, §185) that were offset as emergency spending. Another $29 million for ECP was included in the final appropriation (§753).

[k] "Scorekeeping adjustments" are not necessarily appropriated items and may not be shown in appropriations committee tables but are part of the official CBO score (accounting) of the bill. They predominantly include "negative subsidies" in loan program accounts and adjustments for disaster designations in the bill.

Policy-Related Provisions

In addition to setting budgetary amounts, the Agriculture appropriations bill has also been a vehicle for policy-related provisions that direct how the executive branch should carry out the appropriation. These provisions may have the force of law if they are included in the text of the appropriation, usually in the General Provisions, but their effect is generally limited to the current fiscal year. Sometimes, the provisions amend the U.S. Code and have long-standing effects.

The explanatory statement that accompanies the final appropriation, and the House and Senate report language that accompanies the committee-reported bills, may also provide policy instructions. These documents do not have the force of law but often explain congressional intent, which the agencies are expected to follow.

Table 4 compares some of the major policy provisions that have been identified in the General Provisions (Title VII) of the FY2019 Agriculture appropriations bills. Many of these provisions have been included in past years' appropriations laws.

Table 4. Selected Policy Provisions in FY2019 Agriculture Appropriations

House-reported H.R. 5961	Senate-passed H.R. 6147, Division C
Disaster payments. In general, prohibits the use of Section 32, clause (3), to reestablish farmers' purchasing power by making payments to farmers. However, allows an exception to use up to $350 million of carryover for this purpose. (§715)	Similar to House provision. (§715)
Rural definition. Defines *rural* for the water and waste disposal guaranteed loan program as a city, town, or unincorporated area with no more than 20,000 people. (§728) Directs that eligibility for rural development programs shall not include incarcerated prison populations. (§744)	Similar to House provisions. (§§728, 744)

House-reported H.R. 5961	**Senate-passed H.R. 6147, Division C**
Persistent poverty counties. Requires that at least 10% of the funds in certain rural development programs shall be allocated to persistent poverty counties, defined as any county that has had 20% or more of its population living in poverty over the past 30 years. (§750)	No comparable provision.
Income verification. Provides USDA access to Social Security and Internal Revenue Service data to verify the income of individuals participating in certain rural housing programs. (§747)	Similar to House provision. (§748)
American steel. Prohibits funding for the rural water, wastewater, waste disposal, and solid waste management projects unless all of the iron and steel products are produced in the United States. (§743)	Similar to House provision. (§742)
No comparable provision.	Agriculture risk coverage (ARC) pilot. Directs USDA to conduct a pilot program to make payments for the ARC program that addresses disparities in the calculation of yields among counties. (§747)
No comparable provision.	Industrial hemp. Prohibits funding in contravention of the 2014 farm bill provision for research on industrial hemp or the transportation, processing, sale, or use of industrial hemp or seeds that are grown in accordance with the farm bill provision. (§729)
No comparable provision.	National Bio- and Agro-defense Facility (NBAF). Authorizes ARS and APHIS to appoint up to 50 employees annually in FY2019-2025 for NBAF at a rate of pay that exceeds the General Schedule. (§738)
No comparable provision.	Horse slaughter. Prohibits USDA from conducting horse slaughter inspection. (§758)
Animal Welfare Act. Prohibits funding to issue or renew licenses to class B dealers who sell dogs and cats for use in research, experiments, teaching, or testing. (§741)	No comparable provision.
Lab-grown meat. Directs USDA to regulate products that are made from cells of amenable species of livestock that are grown under controlled conditions for human food to prevent adulteration and misbranding.	No comparable provision.

Table 4. (Continued)

House-reported H.R. 5961	Senate-passed H.R. 6147, Division C
Applies to FY2018 and thereafter. (§736)	
Poultry from China. Prohibits funds to buy raw or processed poultry products from China for the National School Lunch Program, School Breakfast Program, Child and Adult Care Food Program, and Summer Food Service Program. (§748) Prohibits funding to finalize the proposed rule, "Eligibility of the People's Republic of China to Export to the United States Poultry Products," until USDA verifies certain conditions. (§751)	No comparable provisions.
School meal prices. Exempts certain school food authorities (those without a negative balance in their school food service accounts as of December 31, 2018) from paid meal equity requirements in school year 2019-2020. (§757)	No comparable provision.
Vegetables in school breakfast. Prohibits funding to implement or enforce the portion of a School Breakfast Program regulation that limits substituting fruits with certain vegetables. (§767)	No comparable provision.
SNAP retailer standards. Prohibits funding to administer the "variety requirements" in the final rule	Similar to House provision. (§728)
"Enhancing Retailer Standards in SNAP" until the Secretary amends the definition of variety to increase the number of items that qualify as acceptable varieties in each staple food category. (§727)	
SNAP disclosure. Exempts from Freedom of Information Act disclosure SNAP transaction data that contain information specific to a store, store location, person, or other entity. (§768)	No comparable provision.
Human embryos. Prohibits FDA from using funds to accept any investigational new drug application for "research in which a human embryo is intentionally created or modified to include a heritable genetic modification." (§732)	Similar to House provision. (§733)

House-reported H.R. 5961	Senate-passed H.R. 6147, Division C
Partially hydrogenated oils. Directs FDA that no food containing partially hydrogenated oils shall be considered to be adulterated on that basis so long as it enters interstate commerce before June 18, 2018. (§740)	Similar to House provision. (§736)
Sodium. Prevents FDA from developing regulations or issuing final guidance for population-wide sodium reduction actions until a dietary reference intake report is completed. (§752)	Similar to House provision. (§750)
Grapes for wine. Prohibits FDA funds to enforce the final rule "Standards for the Growing, Harvesting, Packing, and Holding of Produce for Human Consumption" with respect to grape varietals used solely for wine. (§755)	Similar to House provision. (§753)
Dietary guidelines. Directs USDA to submit a report to Congress within 180 days about the process used to establish the 2020-2025 Dietary Guidelines for Americans, including the recommendations from the National Academies of Science report, "Redesigning the Process for Establishing the Dietary Guidelines for Americans." (§763)	No comparable provision.
Added sugars. Prohibits funds to administer the final rule "Food Labeling: Revision of the Nutrition and Supplement Facts Labels" to the extent that it requires added sugars labeling for single-ingredient foods to which sugars or sweeteners are not added (e.g., honey or maple syrup). (§764)	Similar to House provision. (§768)
Genetically engineered salmon. Directs that disclosure requirements for genetically engineered salmon or finfish follow the National Bioengineered Food Disclosure Standard and subsequent USDA regulations. (§766)	Prohibits FDA from allowing genetically engineered salmon into interstate commerce until it publishes final labeling guidelines. (§740)
CFTC statutes. Amends the Commodity Exchange Act to address the initial margin requirements for certain swap transactions. (§738) Further amends the act by directing CFTC to consider cost-benefit analysis in its regulations (using language from a House-passed bill, H.R. 238). (§760)	No comparable provision.

Table 4. (Continued)

House-reported H.R. 5961	Senate-passed H.R. 6147, Division C
Cigar regulation. Prohibits funds to administer the FDA final "deeming rule" from May 10, 2016, with respect to traditional large and premium cigars. (§769) Prohibits funds to regulate cigars and pipe tobacco differently than April 25, 2014, the date that such tobacco products were proposed to be deemed subject to certain requirements. (§770)	No comparable provision.
Tobacco products. Amends the Federal Food, Drug, and Cosmetic Act concerning the registration of substantially equivalent tobacco products.	No comparable provision.
(§771)	
Vapor products. Directs the Secretary of Health and Human Services to issue a notice of proposed rulemaking, within 21 months, establishing a product standard for vapor products and to promulgate a final rule within 36 months. (§772) Deems a vapor product to be misbranded if it is advertised in a nonadult publication, specifies terms for distributing vapor products in other than direct face-to-face sales, directs the Secretary of Health and Human Services to promulgate regulations requiring labeling of vapor products, and imposes registration requirements on vapor product retailers. (§773) Amends the Pro-Children Act of 1994 to extend the prohibition on smoking in certain facilities that provide children's services to vapor products. (§774) Amends the Federal Food, Drug, and Cosmetic Act to add age verification requirements for remote sales of vapor products. (§775) Directs the Secretary of Health and Human Services to submit a report to Congress within 180 days about the Youth Vapor Product Education, Prevention, and Enforcement Program. (§776)	No comparable provision.
Electronic nicotine delivery systems (ENDS). Directs the FDA commissioner to conduct a study and submit a report to Congress within 180 days on preventing the use of ENDS by youth. (§777)	No comparable provision.

House-reported H.R. 5961	Senate-passed H.R. 6147, Division C
No comparable provision.	Prevents certain changes from occurring under implementation of the Grain Standards Act. (§759)
No comparable provision.	Provides for technical assistance to respond to volcanic eruptions. (§760)
No comparable provision.	Directs USDA and the National Oceanic and issue a report on ocean farming practices and ways to deacidify oceans. (§761)
No comparable provision.	Directs USDA to submit a report to Congress describing how conservation programs may be used to conserve ocelots (wild cats native to the southwestern United States). (§762)
No comparable provision.	Directs the Rural Housing Service to submit a report to Congress about the housing affected by Title V of the Housing Act of 1949, its loan status, and its affordability. (§763)
No comparable provision.	Directs the FDA to revise the advice in a notice about eating fish, to be consistent with nutrition science recognized by FDA about seafood consumption. (§764)
No comparable provision.	Amends implementation of a broadband program in the 2018 Agriculture Appropriation regarding eligibility determination and expansion efforts. (§765)

Source: CRS.

APPENDIX A. BUDGET SEQUESTRATION

Sequestration is a process to reduce federal spending through automatic, largely across-the-board reductions that permanently cancel mandatory and/or discretionary budget authority.[1] Sequestration is triggered as a budget enforcement mechanism when federal spending would exceed statutory budget goals.[2] Sequestration is currently authorized by the Budget Control Act of 2011 (BCA; P.L. 112-25).

---

[1] CRS Report R43411, *The Budget Control Act of 2011: Legislative Changes to the Law and Their Budgetary Effects.*

[2] CRS Report R42972, *Sequestration as a Budget Enforcement Process: Frequently Asked Questions.*

Table A-1 shows the rates of sequestration that have been announced and the total amounts of budget authority that have been cancelled from accounts in Agriculture appropriations. *Table A-2* provides additional detail at the account level for mandatory accounts.

Table A-1. Sequestration from Accounts in Agriculture Appropriations (budget authority that was sequestered, in millions of dollars)

	Discretionary Accounts		Mandatory Accounts	
Fiscal Year	Rate	Amount	Rate	Amount
2013	5.0%	1,153	5.1%	713
2014	—	—	7.2%	1,052
2015	—	—	7.3%	1,153
2016	—	—	6.8%	1,819
2017	—	—	6.9%	1,686
2018	—	—	6.6%	1,316
2019	—	—	6.2%	1,530

Source: CRS, compiled from OMB, Reports to the Congress on the Joint Committee Reductions, various fiscal years. Available for FY2018-FY2019 at https://www. whitehouse.gov/omb/legislative/sequestration-reports-orders and for FY2013-FY2017 at https://obamawhitehouse.archives.gov/omb/legislative_reports/sequest ration.

Notes: Sequestration rates listed here are for nonexempt, nondefense accounts. Amount totals were computed by CRS as compiled in Table A-2.

Discretionary Spending

For discretionary spending, sequestration is authorized through FY2021 if discretionary defense and nondefense spending exceed caps that are specified in statute (2 U.S.C. 901(c)).

In FY2013, the timing of the appropriations acts and the first year of sequestration resulted in triggering sequestration on discretionary spending.

In FY2014-FY2018, Bipartisan Budget Acts in 2013, 2015, and 2018 (BBAs; P.L. 113-67, P.L. 114-74, and P.L. 115-123, respectively) have avoided sequestration on discretionary spending. These BBAs raised the discretionary budget caps that were placed in statute by the BCA and allowed Congress to enact larger appropriations than would have been allowed.

For FY2019, the BBA in 2018 similarly provides a higher discretionary cap that may avoid sequestration (see "Discretionary Budget Caps and Subcommittee Allocations").

Mandatory Spending

Authorization of Sequestration

For mandatory spending, sequestration is presently authorized through FY2027, having been amended and extended by acts that were subsequent to the BCA (2 U.S.C. 901a(6)). That is, sequestration continues to apply annually to certain accounts of mandatory spending and is not avoided by the BBAs (*Table A-1*).

The original FY2021 sunset on the sequestration of *mandatory* accounts has been extended four times as an offset to pay for avoiding sequestration on *discretionary* spending in the near term or as a general budgetary offset for other authorization acts:

1. Congress extended the duration of mandatory sequestration by two years (until FY2023) as an offset in BBA 2013.[3]
2. Congress extended it by another year (until FY2024) to maintain retirement benefits for certain military personnel (P.L. 113-82).
3. Congress extended sequestration on nonexempt mandatory accounts another year (until FY2025) as an offset in BBA 2015.[4]

---

[3] CBO, *Bipartisan Budget Act of 2013*, December 11, 2013, https://www.cbo.gov/publication/44964.

[4] CBO, *Bipartisan Budget Act of 2015*, October 28, 2015, https://www.cbo.gov/publication/50938. In addition to extending sequestration and related to mandatory spending in agriculture, crop insurance was used as an additional budgetary offset in BBA 2015. The effect was temporary, however, and the crop insurance reduction was restored. For more background, see the section on crop insurance and the Standard Reinsurance Agreement in CRS Report R44240, *Agriculture and Related Agencies: FY2016 Appropriations*.

APPENDIX B. ACTION ON AGRICULTURE APPROPRIATIONS

Table B-1. Congressional Action on Agriculture Appropriations Since FY1997

	House Action			Senate Action			Final Appropriation			
Fiscal Year	Subcmte.	Cmte.	Floor	Subcmte.	Cmte.	Floor	Enacted[a]		Public Law	CRS Report
1997	5/30/1996	6/6/1996	6/12/1996	7/10/1996	7/11/1996	7/24/1996	8/6/1996	E	P.L. 104-180	IB96015
1998	6/25/1997	7/14/1997	7/24/1997	7/15/1997	7/17/1997	7/24/1997	11/18/1997	E	P.L. 105-86	97-201
1999	6/10/1998	6/16/1998	6/24/1998	6/9/1998	6/11/1998	7/16/1998	10/21/1998	O	P.L. 105-277	98-201
2000	5/13/1999	5/24/1999	6/8/1999	6/15/1999	6/17/1999	8/4/1999	10/22/1999	E	P.L. 106-78	RL30201
2001	5/4/2000	5/16/2000	7/11/2000	5/4/2000	5/10/2000	7/20/2000	10/28/2000	E	P.L. 106-387	RL30501
2002	6/6/2001	6/27/2001	7/11/2001	Polled out[b]	7/18/2001	10/25/2001	11/28/2001	E	P.L. 107-76	RL31001
2003	6/26/2002	7/26/2002	—	7/23/2002	7/25/2002	—	2/20/2003	O	P.L. 108-7	RL31301
2004	6/17/2003	7/9/2003	7/14/2003	7/17/2003	11/6/2003	11/6/2003	1/23/2004	O	P.L. 108-199	RL31801
2005	6/14/2004	7/7/2004	7/13/2004	9/8/2004	9/14/2004	—	12/8/2004	O	P.L. 108-447	RL32301
2006	5/16/2005	6/2/2005	6/8/2005	6/21/2005	6/27/2005	9/22/2005	11/10/2005	E	P.L. 109-97	RL32904
2007	5/3/2006	5/9/2006	5/23/2006	6/20/2006	6/22/2006	—	2/15/2007	Y	P.L. 110-5	RL33412
2008	7/12/2007	7/19/2007	8/2/2007	7/17/2007	7/19/2007	—	12/26/2007	O	P.L. 110-161	RL34132
2009	6/19/2008	—	—	Polled out[b]	7/17/2008	—	3/11/2009	O	P.L. 111-8	R40000
2010	6/11/2009	6/18/2009	7/9/2009	Polled out[b]	7/7/2009	8/4/2009	10/21/2009	E	P.L. 111-80	R40721
2011	6/30/2010	—	—	Polled out[b]	7/15/2010	—	4/15/2011	Y	P.L. 112-10	R41475
2012	5/24/2011	5/31/2011	6/16/2011	Polled out[b]	9/7/2011	11/1/2011	11/18/2011	O	P.L. 112-55	R41964
2013	6/6/2012	6/19/2012	—	Polled out[b]	4/26/2012	—	3/26/2013	O	P.L. 113-6	R43110

4. Congress extended sequestration on nonexempt mandatory accounts by another two years (until FY2027) as an offset in BBA 2018 (P.L. 115-123 Division C, §30101(c)).[5]

Exemptions from Sequestration

Some farm bill mandatory programs are exempt from sequestration. Those expressly exempt by statute are the nutrition programs (SNAP, the child nutrition programs, and the Commodity Supplemental Food Program)[6] and the Conservation Reserve Program.[7] Some prior legal obligations in the Federal Crop Insurance Corporation[8] and the farm commodity programs may be exempt[9] as determined by OMB.[10]

Generally speaking, the experience since FY2013 is that OMB has ruled that most of crop insurance is exempt from sequestration, while the farm commodity programs, disaster assistance, and most conservation programs have been subject to it.[11]

Implementation of Sequestration

Sequestration on nonexempt *mandatory* accounts continues in FY2019. Nonexempt mandatory spending is to be reduced by a 6.2% sequestration rate and thus paid at 93.8% of what they would otherwise have provided. This results in a reduction of about $1.5 billion from mandatory agriculture accounts in FY2019.

---

[5] CBO, *Bipartisan Budget Act of 2018*, February 8, 2018, https://www.cbo.gov/ publication/ 53556.

[6] 2 U.S.C. 905 (h).

[7] 2 U.S.C. 905(g)(1)(A).

[8] 2 U.S.C. 905 (g)(2).

[9] 2 U.S.C. 906 (j).

[10] Some administrative expenses may be subject to sequestration, and therefore programs that are otherwise exempt may have a relatively small sequesterable amount compared to their total budget authority as shown, for example, in *Table A-2* for the Federal Crop Insurance Corporation, SNAP, child nutrition programs, and WIC.

[11] CRS Report R42050, *Budget "Sequestration" and Selected Program Exemptions and Special Rules*.

Table A-2. Sequestration of Mandatory Accounts in Agriculture Appropriations (sequestered budget authority in millions of dollars)

	FY2013	FY2014	FY2015	FY2016	FY2017	FY2018	FY2019
Sequestration rate on nonexempt, nondefense mandatory accounts	5.1%	7.2%	7.3%	6.8%	6.9%	6.6%	6.2%
U.S. Department of Agriculture							
Office of the Secretary	—	—	0.9	0.9	0.9	0.9	0.8
Office of Chief Economist	—	—	0.1	0.1	0.1	0.1	0.1
Agricultural Research Service	0.1	0.1	0.1	0.1	0.1	0.1	0.1
National Institute of Food and Agriculture	—	—	—	9.9	10.0	—	—
Extension	0.3	0.4	1.8	—	—	3.3	0.3
Biomass Research and Development	—	—	0.2	0.2	0.2	—	—
Integrated Activities	—	—	7.3	—	—	6.6	5.0
Farm Production and Conservation Business Center	—	—	—	—	—	—	3.7
Animal and Plant Health Inspection Service							
Salaries appropriation	13.6	18.8	21.5	20.1	19.5	18.8	52.1
Miscellaneous Trust Funds	0.1	0.1	0.1	0.1	0.1	0.1	0.1
Food Safety Inspection Service	0.1	0.1	0.1	0.1	0.1	0.1	0.1
Grain Inspection Packers and Stockyards Admin.	2.1	3.0	3.0	3.1	3.1	2.9	—
Agricultural Marketing Service							
Section 32	40.4	79.7	81.9	77.3	79.6	78.1	74.4
Milk Market Orders Assessment Fund	2.9	4.2	4.2	4.0	4.1	4.0	3.8
Perishable Agriculture Commodities Act	0.6	0.8	0.8	0.8	0.8	0.7	0.7
Expenses and refunds	0.4	0.9	0.9	1.3	1.3	0.4	0.4
Payments to states and possessions	—	—	5.3	5.0	5.0	5.6	5.3
Marketing services	—	—	2.2	2.0	2.1	2.0	3.0
Risk Management Agency	—	—	—	—	0.6	—	—
Federal Crop Insurance Corporation	3.0	4.2	5.9	3.5	3.9	3.7	5.0
Farm Service Agency							
Commodity Credit Corporation Fund	329.5	573.7	710.8	1,388.6	1,238.6	904.1	1,088.2
Agricultural Credit Insurance Corporation	—	—	0.1	0.1	0.1	0.1	0.1
Commodity Credit Corporation Export Loans	—	—	—	0.4	0.4	0.3	0.3
Pima Cotton Trust Fund	—	—	—	1.1	1.1	1.1	—
Wool Apparel Manufacturers Trust Fund	—	—	—	2.0	2.1	2.0	1.9
Agricultural Disaster Relief Fund	70.0	—	—	—	—	—	—

Table A-2. (Continued)

	FY2013	FY2014	FY2015	FY2016	FY2017	FY2018	FY2019
Sequestration rate on nonexempt, nondefense mandatory accounts	5.1%	7.2%	7.3%	6.8%	6.9%	6.6%	6.2%
Tobacco Trust Fund	49.0	69.1	—	—	—	—	—
Natural Resources Conservation Service							
Farm bill conservation programs	171.2	263.1	269.9	265.7	281.5	254.9	262.4
Watershed Rehabilitation Program	—	11.9	11.2	4.7	4.7	4.4	3.0
Damage Assessment Restoration Revolving Fund	—	—	—	—	—	—	0.1
Rural Development, Rural Business Coop. Service							
Rural Energy for America Program	1.1	3.0	3.7	3.4	3.5	3.3	3.1
Economic Development Grants	—	—	—	—	—	—	0.9
Rural Microenterprise Investment Program	—	0.2	0.2	0.2	0.2	0.2	—
Energy Assistance Payments	3.3	3.2	1.1	1.0	1.0	1.0	—
Biorefinery Assistance Program	—	—	3.7	3.4	1.4	1.3	—
Rural Economic Development Grants	—	—	—	1.6	—	—	—
Foreign Agricultural Service	0.1	0.1	0.1	0.1	0.1	0.1	0.1
Food and Nutrition Service[a]							
SNAP	4.7	8.0	8.4	9.8	10.6	9.6	9.2
Child Nutrition Programs	2.5	4.2	4.2	3.9	4.3	4.0	3.9
Commodity Assistance Program	1.1	1.5	1.5	1.4	1.4	1.4	1.3
WIC	0.1	0.1	0.1	0.1	0.1	0.1	0.1
Related Agencies							
Food and Drug Administration							
Revolving Fund for Certification	0.4	0.6	0.6	0.6	0.6	0.6	0.6
User Fees	16.3	—	—	—	—	—	—
Farm Credit System Insurance Corporation	0.2	0.3	0.3	0.3	0.3	0.3	0.2
Commodity Futures Trading Commission	0.7	0.9	1.0	2.2	2.2	0.1	0.3
Total	713.3	1,051.9	1,153.0	1,818.9	1,685.6	1,315.9	1,530.3

Source: CRS, compiled from OMB, Reports to the Congress on the Joint Committee Reductions, various fiscal years. Available for FY2018-FY2019 at https://www.whitehouse.gov/omb/ legislative/sequestration-reports-orders and for FY2013-FY2017 at https://obamawhitehouse. archives.gov/omb/legislative_reports/sequestration.

Notes: Sequestration rates are for nonexempt, nondefense accounts. The sequesterable budget authority for each cell may be computed by dividing the amount of sequestration by the sequestration rate. Column totals were computed by CRS based on the jurisdiction of the House Agriculture Appropriations Subcommittee.

[a] Benefits from the nutrition programs are generally exempt from sequestration by statute, but some administrative expenses in these programs may be subject to sequestration, and therefore a relatively small portion of the total budget authority may be sequesterable.

APPENDIX B. ACTION ON AGRICULTURE APPROPRIATIONS, FY1997-FY2019

Table B-1. Congressional Action on Agriculture Appropriations Since FY1997

House Action				Senate Action			Final Appropriation			
Fiscal Year	Subcmte.	Cmte.	Floor	Subcmte.	Cmte.	Floor	Enacted[a]		Public Law	CRS Report
1997	5/30/1996	6/6/1996	6/12/1996	7/10/1996	7/11/1996	7/24/1996	8/6/1996	E	P.L. 104-180	IB96015
1998	6/25/1997	7/14/1997	7/24/1997	7/15/1997	7/17/1997	7/24/1997	11/18/1997	E	P.L. 105-86	97-201
1999	6/10/1998	6/16/1998	6/24/1998	6/9/1998	6/11/1998	7/16/1998	10/21/1998	O	P.L. 105-277	98-201
2000	5/13/1999	5/24/1999	6/8/1999	6/15/1999	6/17/1999	8/4/1999	10/22/1999	E	P.L. 106-78	RL30201
2001	5/4/2000	5/16/2000	7/11/2000	5/4/2000	5/10/2000	7/20/2000	10/28/2000	E	P.L. 106-387	RL30501
2002	6/6/2001	6/27/2001	7/11/2001	Polled outb	7/18/2001	10/25/2001	11/28/2001	E	P.L. 107-76	RL31001
2003	6/26/2002	7/26/2002	—	7/23/2002	7/25/2002	—	2/20/2003	O	P.L. 108-7	RL31301
2004	6/17/2003	7/9/2003	7/14/2003	7/17/2003	11/6/2003	11/6/2003	1/23/2004	O	P.L. 108-199	RL31801
2005	6/14/2004	7/7/2004	7/13/2004	9/8/2004	9/14/2004	—	12/8/2004	O	P.L. 108-447	RL32301
2006	5/16/2005	6/2/2005	6/8/2005	6/21/2005	6/27/2005	9/22/2005	11/10/2005	E	P.L. 109-97	RL32904
2007	5/3/2006	5/9/2006	5/23/2006	6/20/2006	6/22/2006	—	2/15/2007	Y	P.L. 110-5	RL33412
2008	7/12/2007	7/19/2007	8/2/2007	7/17/2007	7/19/2007	—	12/26/2007	O	P.L. 110-161	RL34132
2009	6/19/2008	—	—	Polled out[b]	7/17/2008	—	3/11/2009	O	P.L. 111-8	R40000
2010	6/11/2009	6/18/2009	7/9/2009	Polled out[b]	7/7/2009	8/4/2009	10/21/2009	E	P.L. 111-80	R40721
2011	6/30/2010	—	—	Polled out[b]	7/15/2010	—	4/15/2011	Y	P.L. 112-10	R41475
2012	5/24/2011	5/31/2011	6/16/2011	Polled out[b]	9/7/2011	11/1/2011	11/18/2011	O	P.L. 112-55	R41964
2013	6/6/2012	6/19/2012	—	Polled out[b]	4/26/2012	—	3/26/2013	O	P.L. 113-6	R43110
2014	6/5/2013	6/13/2013	—	6/18/2013	6/20/2013	—	1/17/2014	O	P.L. 113-76	R43110
2015	5/20/2014	5/29/2014	—	5/20/2014	5/22/2014	—	12/16/2014	O	P.L. 113-235	R43669

Table B-1 (Continued)

House Action				Senate Action			Final Appropriation			
Fiscal Year	Subcmte.	Cmte.	Fiscal Year	Subcmte.	Cmte.	Fiscal Year	Subcmte.	Cmte.	Fiscal Year	Subcmte.
2016	6/18/2015	7/8/2015	—	7/14/2015	7/16/2015	—	12/18/2015	O	P.L. 114-113	R44240
2017	4/13/2016	4/19/2016	—	5/17/2016	5/19/2016	—	5/5/2017	O	P.L. 115-31	R44588
2018	6/28/2017	7/12/2017	9/14/2017	7/18/2017	7/20/2017	—	3/23/2018	O	P.L. 115-141	R45128
2019	5/9/2018	5/16/2018	—	5/22/2018	5/24/2018	8/1/2018	—	—	—	R45230

Source: CRS.

[a] E = Enacted as standalone appropriation (seven times over 22 years); O = Omnibus appropriation (1 3 times); Y = Year-long CR (two times).

[b] A procedure that permits a Senate subcommittee to transmit a bill to its full committee without a formal markup session. See CRS Report RS22952, Proxy Voting and Polling in Senate Committee.

In: Agricultural Policy …
Editor: Albertus Brinkhuis

ISBN: 978-1-53615-215-9

Chapter 3

AGRICULTURE AND RELATED AGENCIES: FY2018 APPROPRIATIONS*

Congressional Research Service

SUMMARY

The Agriculture appropriations bill funds the U.S. Department of Agriculture (USDA) except for the Forest Service. It also funds the Food and Drug Administration (FDA) and—in even-numbered fiscal years—the Commodity Futures Trading Commission (CFTC).

Agriculture appropriations include both mandatory and discretionary spending. Discretionary amounts, though, are the primary focus during the bill's development, since mandatory amounts are generally set by authorizing laws such as the farm bill.

The largest discretionary spending items are the Special Supplemental Nutrition Program for Women, Infants, and Children (WIC); agricultural research; FDA; rural development; foreign food aid and trade; farm assistance programs; food safety inspection; conservation; and animal and plant health programs. The main mandatory spending

* This is an edited, reformatted and augmented version of Congressional Research Service Publication No. R45128, Updated March 23, 2018.

items are the Supplemental Nutrition Assistance Program (SNAP), child nutrition, crop insurance, and the farm commodity and conservation programs paid by the Commodity Credit Corporation.

Congress passed the FY2018 Consolidated Appropriations Act on March 23, 2018 (P.L. 115-141). The discretionary total of the FY2018 Agriculture appropriation (Division A) is $23.3 billion. This is an increase of $2.1 billion above the amount enacted in FY2017 (+10%). The appropriations also carry nearly $123 billion of mandatory spending, though that amount is largely determined in separate authorizing laws. Thus, the overall total of the enacted FY2018 Agriculture appropriation is about $146 billion.

Compared to FY2017, the enacted appropriation increases spending primarily through an extra $1.38 billion in General Provisions, including an additional $500 million for rural water programs, $600 million for rural broadband, $116 million for Food for Peace foreign food aid, and $94 million for opioid enforcement and surveillance.

The Consolidated Appropriations Act also increases regular funding for agricultural research by $138 million, and Food for Peace by $134 million. Unlike the past decade, the FY2018 Agriculture appropriation does not include as many changes to mandatory program spending (CHIMPS), such as to the Environmental Quality Incentives Program (EQIP).

Both the House and Senate Appropriations Committees reported Agriculture appropriations bills in July 2017 (H.R. 3268, S. 1603). As the beginning of the fiscal year approached, the House passed a consolidated bill in September 2017 that included an agriculture appropriation (Division B of H.R. 3354). The full Senate did not consider an agriculture appropriations bill on the floor.

The government continued to operate for the first six months of the fiscal year under continuing resolutions (CRs). The last CR, the Bipartisan Budget Act of 2018 (P.L. 115-123), also enacted supplemental appropriations that included agricultural assistance, amended certain farm bill statutes, and passed new, higher budget caps that facilitated the final appropriation.

STATUS OF FY2018 AGRICULTURE APPROPRIATIONS

Congress passed the FY2018 Consolidated Appropriations Act on March 23, 2018 (P.L. 115-141). Both the House and Senate Appropriations Committees had reported Agriculture appropriations bills for FY2018

(H.R. 3268, S. 1603). The House also had passed a consolidated bill that included agriculture (H.R. 3354). The full Senate did not consider the Agriculture appropriations bill on the floor. New, higher budget caps in the Bipartisan Budget Act of 2018 (P.L. 115-123) facilitated the final appropriation.

Specifically, the House Appropriations Committee passed H.R. 3268 on July 12, 2017, and the Senate Appropriations Committee passed S. 1603 on July 20, 2017. On September 14, the House passed an eight-bill consolidated appropriation, H.R. 3354, with the Agriculture bill as Division B that included amendments to the House-reported version (*Table 1, Figure 1, Appendix B*).

Table 1. Legislative Status of FY2018 Agriculture Appropriations

House Action			**Senate Action**			**Final Appropriation**	
Subcmte.	Cmte.	Floor	Subcmte.	Cmte.	Floor	Enacted	Public Law
6/28/2017 Draft Voice vote	7/12/2017 H.R. 3268 H.Rept. 115-232 Voice vote	9/14/2017 H.R. 3354 (Division B) Vote of 211-198	7/18/2017 Voice vote	7/20/2017 S. 1603 S.Rept. 115-131 Vote of 31-0	—	3/23/2018 Votes H: 256-167 S: 65-32	P.L. 115-141 Text and explanation[a]

The discretionary total of the enacted Agriculture appropriation is \$23.3 billion, which is \$2.1 billion more than enacted in FY2017 (+10%), on a comparable basis that includes the Commodity Futures Trading Commission (CFTC; *Table 2*).[1] The Administration had proposed a \$15.8 billion discretionary total (a reduction of 25%), the House-passed bill \$20 billion, and the Senate-reported bill \$20.53 billion.

[1] Jurisdiction for CFTC appropriations differs between the chambers. Since FY2008, CFTC is marked up in the Agriculture Subcommittee of the House Appropriations Committee and in the Financial Services and General Government Subcommittee of the Senate Appropriations Committee. The enacted CFTC appropriation is carried in the Agriculture bill in even-numbered fiscal years and in the Financial Services bill in odd-numbered fiscal years.

The appropriations also carry mandatory spending—though that is largely determined in separate authorizing laws—that total nearly $123 billion. Thus, the overall total of the enacted FY2018 Agriculture appropriation is $146 billion.

The Trump Administration released its full FY2018 budget request on May 23, 2017,[2] along with the detailed justification from the U.S. Department of Agriculture (USDA).[3]

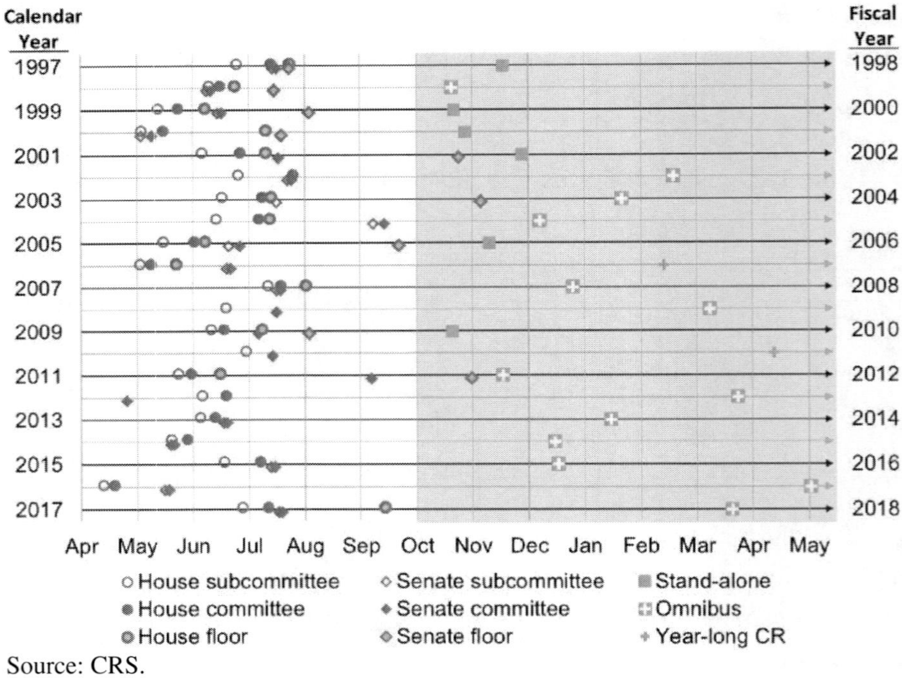

Source: CRS.

Figure 1. Timeline of Action on Agriculture Appropriations, FY1998-FY2018.

[2] Office of Management and Budget (OMB), *FY2018 Budget of the U.S. Government*, especially in the Appendix, http://www.whitehouse.gov/omb/budget/Appendix.

[3] USDA, *FY2018 USDA Budget Summary*; and USDA, *2018 Explanatory Notes*, http://www.obpa.usda.gov.

Table 2. Agriculture and Related Agencies Appropriations, by Title, FY2017-FY2018 (budget authority in millions of dollars)

Title of Agriculture Appropriations Act	**FY2017**	**FY2018**				**Change FY2017 to FY2018 Enacted**	
	P.L. 115-31	Admin. Request	House H.R. 3354	S. Cmte. S. 1603	P.L. 115-141		
Agricultural Programs: Discretionary[a]	7,107.7[a]	4,861.2	5,141.2	5,363.1	5,626.8	+219.0b	+4.0%[b]
Mandatory (Section 32 in FY2018)[a]	*31,280.2*[a]	*1,344.0*	*1,344.0*	*1,344.0*	*1,344.0*	*+22.0b*	*+1.7%*[b]
Subtotal[a]	*38,387.9*[a]	*6,205.2*	*6,485.2*	*6,707.1*	*6,970.8*	*+240.5b*	*+3.6%*[b]
Farm Production and Conservation	1,027.4[a]	2,330.2	2,527.8	2,726.4	2,735.6	+9.8b	+0.4%[b]
Programs[a]							
Mandatory (CCC, crop insurance in FY2018)[a]	—[a]	*25,728.5*	*25,728.5*	*25,728.5*	*23,198.3*	*-6,759.9*[b]	*-22.6%*[b]
Subtotal[a]	—[a]	*28,058.7*	*28,256.3*	*28,454.9*	*25,933.9*	*-6,750.0*[b]	*-20.7%*[b]
Rural Development	3,069.2	2,157.5	2,807.1	2,949.8	3,000.9[c]	-68.3	-2.2%
Domestic Food Programs: Discretionary	6,884.7	6,615.9	6,664.5	6,887.9	6,709.8	-174.9	-2.5%
Mandatory (SNAP and child nutrition)	*101,226.7*	*97,845.8*	*97,842.9*	*97,842.9*	*98,209.6*	*-3,017.0*	*-3.0%*
Subtotal	*108,111.3*	*104,461.7*	*104,507.4*	*104,730.8*	*104,919.4*	*-3,191.9*	*-3.0%*
Foreign Assistance	1,872.9[c]	195.1	1,804.7	2,013.7	2,021.0[c]	+148.1	+7.9%
Related Agencies: Food and Drug Admin.	2,771.2	1,828.5	2,768.1	2,772.2	2,811.9[c]	+40.7	+1.5%
Commodity Futures Trading Commission	[250.0]	250.0	248.0	[250.0][d]	249.0	-1.0	-0.4%
General Provisions: CHIMPSe & rescissions	-1,598.0	-1,931.7	-1,490.0	-1,741.0	-801.0	+797.0	-49.9%
Disaster/emergency programs	234.8	—	—	—	—	-234.8	-100%
Other appropriations	237.4	—	6.5	35.1	1,378.1	+1,140.7	+480%
Scorekeeping adjustments[f]	-524.0	-485.0	-481.0	-482.0	-481.0	+43.0	-8.2%
Subtract disaster declaration in this bill	-206.1	—	—	—	—	+206.1	-100%
Discretionary: Senate basis w/o CFTC	20,877.0	15,571.6	[19,748.9]	20,525.1	[23,002.0]	+2,125.0	+10.2%
Discretionary: House basis w/ CFTC	[21,127.0]	15,821.6	19,996.9	[20,775.1]	23,251.0	+2,124.0	+10.1%

Table 2. (Continued)

Title of Agriculture Appropriations Act	FY2017		FY2018			Change FY2017 to FY2018 Enacted	
	P.L. 115-31	Admin. Request	House H.R. 3354	S. Cmte. S. 1603	P.L. 115-141		
Mandatory	*132,506.9*	*124,918.3*	*124,915.4*	*124,915.4*	*122,752.0*	*-9,754.9*	*-7.4%*
Total: Senate basis w/o CFTC	*153,383.9*	*140,489.9*	*144,664.3*	*145,440.5*	*145,754.0*	*-7,629.9*	*-5.0%*
Total: House basis w/ CFTC	*153,633.9*	*140,739.9*	*144,912.3*	*145,690.5*	*146,003.0*	*-7,630.9*	*-5.0%*

Source: CRS, using appropriations text and reports, and unpublished CBO tables. House column reflects H.R. 3354 as amended from H.R. 3268

Notes: Amounts are nominal discretionary budget authority in millions of dollars unless labeled otherwise.

Excludes amounts in supplemental appropriations acts. Bracketed amounts are not in the official totals due to differing House-Senate jurisdiction for CFTC.

[a] Row headings reflect recent USDA reorganization. For FY2018, the Farm Service Agency and Risk Management Agency were moved from Title I to Title II, as was Commodity Credit Corporation (CCC) and Federal Crop Insurance Corporation within mandatory spending.

[b] Differences were computed using amounts for FY2017 that were adjusted to reflect the reorganization.

[c] Excludes a portion of the other appropriations that are provided separately in General Provisions.

[d] Senate Financial Services Appropriations Subcommittee, chairman's draft, in lieu of subcommittee markup.

[e] Changes in Mandatory Program Spending (CHIMPS) are reductions made to mandatory programs.

[f] "Scorekeeping adjustments" are not necessarily appropriated but are part of the official CBO accounting.

Scope of Agriculture Appropriations

The Agriculture appropriations bill—formally known as the Agriculture, Rural Development, Food and Drug Administration, and Related Agencies Appropriations Act—funds all of USDA, excluding the U.S. Forest Service. It also funds the Food and Drug Administration (FDA) in the Department of Health and Human Services. In even-numbered fiscal years, the act carries CFTC funding under a practice started in FY2008 for handling House-Senate jurisdictional differences.

Jurisdiction is with the House and Senate Committees on Appropriations and their respective Subcommittees on Agriculture, Rural Development, Food and Drug Administration, and Related Agencies. The bill includes mandatory and discretionary spending, but the discretionary amounts are the primary focus during the bill's development. The scope of the bill is shown in *Figure 2*.

The federal budget process treats discretionary and mandatory spending differently.[1]

- Discretionary spending is controlled by annual appropriations acts and receives most of the attention during the appropriations process. The annual budget resolution[2] process sets spending limits for discretionary appropriations. Agency operations (salaries and expenses) and many grant programs are discretionary.
- Mandatory spending[3]—though carried in the appropriation and usually advanced unchanged—is controlled by budget rules (e.g., PAYGO) during the authorization process.[4] Spending for so-called

---

[1] See CRS Report R44582, *Overview of Funding Mechanisms in the Federal Budget Process, and Selected Examples.*

[2] See CRS Report R42388, *The Congressional Appropriations Process: An Introduction.*

[3] Mandatory spending creates funding stability and consistency compared to annual discretionary appropriations. In agriculture, it originated with farm commodity programs that had uncertain outlays due to weather and the market.

[4] See CRS Report 98-560, *Baselines and Scorekeeping in the Federal Budget Process.*

entitlement programs is set in laws such as the 2014 farm bill[5] and 2010 child nutrition reauthorizations.[6]

In FY2018 Agriculture appropriations, discretionary appropriations were 16% ($23.3 billion) of the total. Mandatory spending carried in the act comprised $123 billion, about 84% of the $146 billion total. About $99 billion of the $123 billion mandatory amount could be attributed to programs in the 2014 farm bill (*Figure 2*). Other programs are not in the authorizing jurisdiction of the Agriculture Committees.

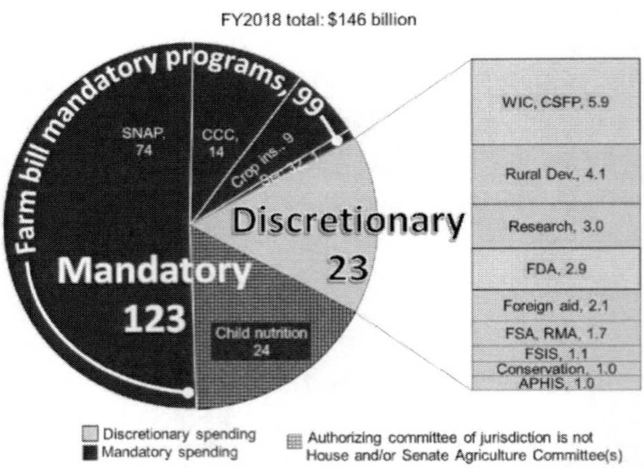

Source: CRS. Does not show some agencies under $0.5 billion. Includes General Provisions with agency.

Notes: SNAP = Supplemental Nutrition Assistance Program; CCC = Commodity Credit Corporation; FCIC = Federal Crop Insurance Corporation; Section 32 = Funds for Strengthening Markets, Income and Supply; WIC = Special Supplemental Nutrition Program for Women, Infants, and Children; CSFP = Commodity Supplemental Food Program; FDA = Food and Drug Administration; FSA = Farm Service Agency; RMA = Risk Management Agency; FSIS = Food Safety and Inspection Service; APHIS = Animal and Plant Health Inspection Service.

Figure 2. Scope of Agriculture and Related Agencies Appropriations. (FY2018 budget authority in billions of dollars)

---

[5] P.L. 113-79. See CRS In Focus IF10783, *Farm Bill Primer: Budget Issues*.

[6] P.L. 111-296. See CRS Report R44373, *Tracking the Next Child Nutrition Reauthorization: An Overview*.

Key Budget Terms

Budget authority is the main purpose of an appropriations act or a law authorizing mandatory spending. It provides the legal basis from which to obligate funds. It expires at the end of a period and is usually available for one year unless specified otherwise (e.g., two years or indefinite). Most amounts in this report are budget authority.

Obligations are contractual agreements between an agency and its clients or employees. They occur when an agency proceeds to spend money from its budget authority. The Antideficiency Act prohibits agencies from obligating more budget authority than is provided in law, such as during a government shutdown.

Outlays are the payments (cash disbursements) that satisfy a valid obligation. Outlays may differ from budget authority or obligations because payments from an agency may not occur until services are fulfilled, goods are delivered, or construction is completed, even though an obligation occurred.

Program level represents the sum of the activities supported or undertaken by an agency. A program level may be higher than a budget authority if the program (1) receives *user fees* that can be used to pay for activities, (2) makes or guarantees **loans** that are leveraged on the expectation of repayment (more than $1 of loan authority for $1 of budget authority), or (3) receives *transfers* from other agencies.

Rescissions are adjustments that cancel or reduce budget authority after it has been enacted. They score budgetary savings.

CHIMPS (Changes in Mandatory Program Spending) are adjustments in an appropriations act to mandatory budget authority. CHIMPS in appropriations usually reduce or limit spending by mandatory programs for one year and score budgetary savings. They do not change the underlying authority of the program in statute.

Within the discretionary total, the largest discretionary spending items are for the Special Supplemental Nutrition Program for Women, Infants, and Children (WIC); rural development; agricultural research; FDA; foreign food aid and trade; farm assistance program salaries and loans; food safety inspection; conservation; and animal and plant health programs (*Figure 2*).

The main mandatory spending items are the Supplemental Nutrition Assistance Program (SNAP, and other food and nutrition act programs), child nutrition (school lunch and related programs), crop insurance, and farm commodity and conservation programs paid through USDA's Commodity Credit Corporation (CCC). SNAP is referred to as an "appropriated entitlement" and requires an annual appropriation.[7] Amounts for the nutrition program are based on projected spending needs. In contrast, the CCC operates on a line of credit. The annual appropriation provides funding to reimburse the Treasury for using this line of credit.

RECENT TRENDS IN AGRICULTURE APPROPRIATIONS

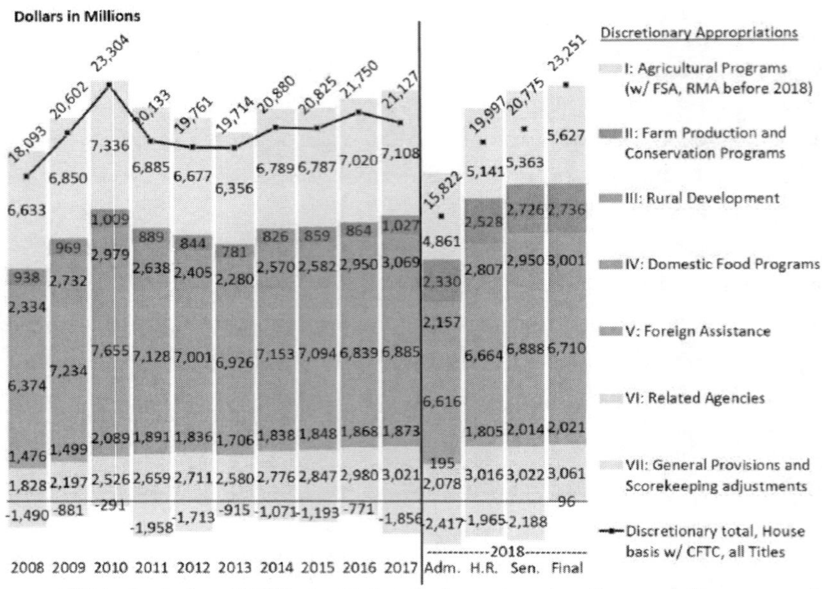

Source: CRS. Includes CFTC in Related Agencies in all years for comparison, regardless of jurisdiction.

Figure 3. Discretionary Agriculture Appropriations, by Title, FY2008-2018.

[7] See CRS Report RS20129, *Entitlements and Appropriated Entitlements in the Federal Budget Process.*

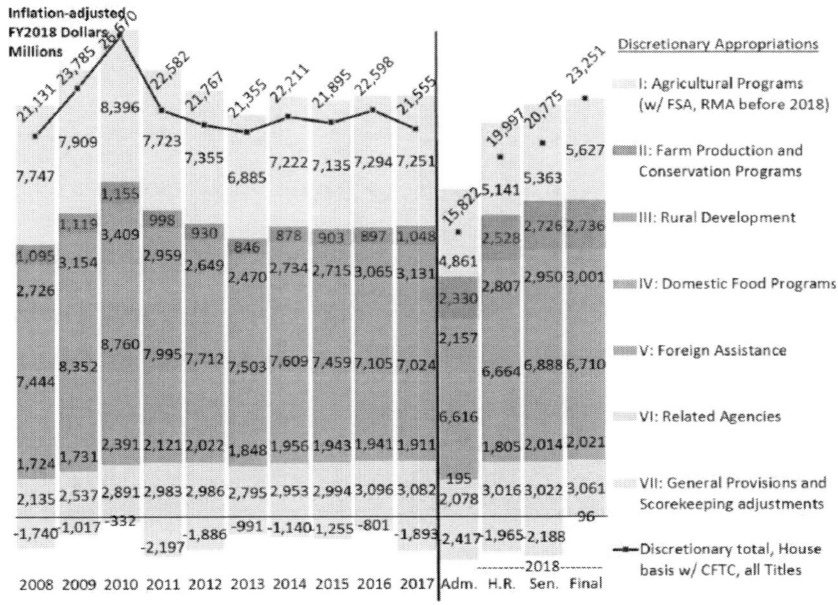

Source: CRS. Includes CFTC in Related Agencies in all years for comparison, regardless of jurisdiction.

Notes: Budget authority adjusted for inflation by CRS using the gross domestic product price deflator.

Figure 4. Inflation-Adjusted Discretionary Agriculture Appropriations, FY2008-2018.

Over time, changes by title of the Agriculture appropriations bill have generally been proportionate to changes in the bill's total discretionary limit, though some activities have sustained relative increases and decreases. Agriculture appropriations peaked in FY2010, declined through FY2013, and since then have increased (*Figure 3*). Comparisons to historical benchmarks, though, may depend on adjustments for inflation and other factors (*Figure 4*).

The stacked bars in *Figure 3* represent the discretionary spending that was authorized for each appropriations title since FY2008. Prior to FY2018 enactment, the total of the positive stacked bars is the budget authority contained in Titles I-VI. It was higher than the official discretionary spending allocation for the subcommittee (the line) because of the budgetary offset from negative amounts in Title VII general provisions and other scorekeeping adjustments. General provisions were net negative mostly because of rescissions and limits placed on mandatory programs.

ACTION ON FY2018 APPROPRIATIONS

Administration's Budget Request

Like most new Administrations, the Trump Administration released its FY2018 budget request in 2017 later than the usually expected first week of February.[8] It released an outline (sometimes called a "skinny budget")[9] on March 16, 2017, that expressed intentions primarily at the Cabinet level. For USDA, it proposed a 21% reduction, including eliminating funding for some programs.

The Administration's budget outline overlapped with Congress finishing FY2017 appropriations. On May 5, 2017, the explanatory statement for the FY2017 appropriation (P.L. 115-31) addressed the direction of the new request for FY2018 by reminding the Administration of Congress's role:

> USDA and FDA should be mindful of Congressional authority to determine and set final funding levels for fiscal year 2018. Therefore, the agencies should not presuppose program funding outcomes and prematurely initiate action to redirect staffing prior to knowing final outcomes on fiscal year 2018 program funding.[10]

On May 23, 2017, the Administration released its full FY2018 budget request.[11] USDA concurrently released its full budget summary and

---

[8] See CRS Report R42388, The Congressional Appropriations Process: An Introduction.

[9] The White House, America First: A Budget Blueprint to Make America Great Again, March 16, 2017.

[10] Congressional Record, May 3, 2017, p. H3328.

[11] OMB, FY2018 Budget of the U.S. Government, Appendix.

justification,[12] as did the FDA.[13] Some OMB proposals[14] were not consistent with the USDA request. As an independent agency, CFTC requested a different amount in its budget justification[15] than the Administration requested.

For Agriculture appropriations (including CFTC), the Trump Administration requested $15.82 billion of discretionary spending, $5.3 billion less than in FY2017 (-25%; *Table 2, Figure 3*).

From these documents, the congressional appropriations committees evaluated the request and began to consider their own bills in the summer of 2017.

House Action

At the time the House appropriations subcommittees began marking up their FY2018 bills, an FY2018 budget resolution had not yet been adopted. Therefore, the House Appropriations Committee incrementally made "302(b)" allocations[16] to its subcommittees to facilitate markups beginning in June 2017.[17] The Agriculture Subcommittee was allowed nearly $20 billion of discretionary budget authority for FY2018.

---

[12] USDA, FY2018 USDA Budget Summary; and USDA, 2018 Budget Explanatory Notes, http://www.obpa.usda.gov.

[13] FDA, FY2018 FDA Justification of Estimates for Appropriations, https://www.hhs.gov/about/budget/index.html.

[14] OMB, FY2018 Budget of the U.S. Government, Major Savings and Reforms.

[15] CFTC, *FY2018 CFTC Budget Request*, http://www.cftc.gov/About/CFTCReports/ssLINK/cftcbudget2018.

[16] Budget enforcement for appropriations has both procedural and statutory elements. The procedural elements are associated with the budget resolution and are enforced through points of order. The appropriations committees and subcommittees receive procedural limits on budget authority, referred to as 302(a) and 302(b) allocations, respectively (see CRS Report R42388, *The Congressional Appropriations Process: An Introduction*). The statutory elements impose limits on discretionary spending in FY2012-FY2021 and are enforced through discretionary budget caps and sequestration (see CRS Report R42972, *Sequestration as a Budget Enforcement Process: Frequently Asked Questions*).

[17] For example, for the limit for Agriculture appropriations, see House Appropriations Committee, "Revised Interim Suballocation of Budget Allocations for FY2018," https://appropriations.house.gov/uploadedfiles/sbdv-3.pdf.

The House Agriculture Appropriations Subcommittee marked up its FY2018 bill on June 28, 2017, by voice vote.[18] On July 12, 2017, the full Appropriations Committee passed and reported an amended bill (H.R. 3268, H.Rept. 115-232) by voice vote (*Table 1, Figure 1*).

As the beginning of the fiscal year neared without many floor-passed appropriations bills, the House passed on September 14, 2017, an eight-bill consolidated appropriation (H.R. 3354) with Agriculture as Division B. It included several budget-neutral amendments to the reported version.

The $20 billion discretionary total in the House-passed FY2018 Agriculture appropriation (*Table 2, Figure 3*) was officially $1.13 billion less than enacted in FY2017 (-5.3%, on a comparable basis that adds CFTC back to the FY2017 Agriculture appropriation's total).

Compared to FY2017, the House-passed bill would have achieved this reduction primarily by reducing rural development by $262 million, nutrition assistance by $220 million, farm and conservation programs by $199 million, department administration by $123 million, and research by $98 million, among other changes. *Table 3* provides details at the agency level.

Senate Action

In the absence of an adopted FY2018 budget resolution, the Senate Appropriations Committee made 302(b) allocations to its subcommittees in July 2017, similar to the House approach.[19] The Agriculture Subcommittee was allowed $20.53 billion of discretionary authority for FY2018.

The Senate Agriculture Appropriations Subcommittee marked up its FY2018 bill on July 18, 2017, by voice vote. Two days later, the full Appropriations Committee passed and reported its amended bill (S. 1603, S.Rept. 115-131) on July 20, 2017, by a vote of 31-0 (*Table 1, Figure 1*).

---

[18] The House subcommittee draft is available at https://appropriations.house.gov/uploadedfiles/bills-115hr-sc-apfy2018-agriculture-agriculture.pdf.

[19] For example, see "FY2018 Funding Guidance" at https://www.appropriations. senate.gov/imo/media/doc/072017%20FY2018%20Funding%20Guidance%20-%20Background.pdf.

The discretionary total of the Senate-reported bill was $20.53 billion (*Table 2, Figure 3*), which was $352 million less than enacted in FY2017 without CFTC (-1.7%). The Senate-reported bill would have provided $776 million more than the House-passed bill on a comparable basis without the CFTC.

The Senate-reported bill would have made fewer and smaller reductions compared to FY2017 than the House-passed bill. Compared to FY2017, it would have reduced rural development by $119 million and agricultural research by $57 million and increased foreign food aid by $140 million, providing more for each of these program areas than the House-passed bill (*Table 3*). It would also have made $220 million more in reductions—through rescissions and changing mandatory programs—than the House-passed bill.

Continuing Resolutions

FY2018 began on October 1, 2017, without an enacted appropriation. As a result, Congress has passed CRs to continue to fund the government.[20] In general, a CR continues the funding rate and other provisions of the previous year's appropriation.[21] However, the Office of Management and Budget (OMB) prorates funding to the agencies on an annualized basis for the duration of the CR through a process known as apportionment.[22] CRs may also provide a different amount through anomalies or make specific administrative changes.[23]

1. The first CR for FY2018 (Division D of P.L. 115-56) lasted until December 8, 2017. It continued FY2017 funding levels and

---

[20] CRS Report R42647, Continuing Resolutions: Overview of Components and Recent Practices.

[21] CRS Report RL34700, Interim Continuing Resolutions (CRs): Potential Impacts on Agency Operations.

[22] For example, if a CR lasts for three months, OMB may apportion 3/12 of the previous fiscal year amount during the CR to limit agency spending.

[23] CRS Report R44978, Overview of Continuing Appropriations for FY2018 (P.L. 115-56).

provisions with two general exceptions and two anomalies for the agriculture appropriation:

- A 0.6791% across-the-board reduction (Section 101(b)).
- Sufficient funding to maintain mandatory program levels, including for nutrition programs (Section 111).
- An increase of about $2 million for the Commodity Supplemental Food Program, a domestic food assistance program that predominantly serves the low-income elderly. Rather than the FY2017 funding level of $236 million, the CR provides about $238 million for this program. This anomaly is typically included to maintain caseload and participation based on food costs (Section 116).
- A technical correction for the computation of a rescission to Section 32 funds in light of the availability that is allowed for carryover funds, especially for disaster payments that are at the discretion of the USDA (Section 117).

2. A second CR (P.L. 115-90) extended the provisions and anomalies of the first CR to December 22, 2017.
3. A third CR (P.L. 115-96) extended the CR to January 19, 2018. It also waived PAYGO rules (Section 5002) for the Tax Cuts and Jobs Act (P.L. 115-97) that could have caused a sequestration of mandatory spending as an automatic budgetary offset, which could have affected the farm bill budget.[24]
4. In the absence of a further CR or appropriation by January 19, a three-day government shutdown occurred through January 22, when a fourth CR (P.L. 115- 120) extended the CR to February 8, 2018.

In the absence of a further CR or appropriation by February 8, an overnight government shutdown occurred during the early morning of February 9, when a fifth CR (P.L. 115-123) extended the CR to March 23, 2018. This CR was part of the Bipartisan Budget Act of 2018 that raised

---

[24] CBO, "Effects of Legislation That Would Raise Deficits by an Estimated $1.5 Trillion over the 2018-2027 Period," letter to Steny Hoyer, November 14, 2017.

Table 3. (Continued)

Agency or Major Program	FY2015	FY2016	FY2017	FY2018				Change FY2017 to FY2018 Enacted	
	P.L. 113-235	P.L. 114-113	P.L. 115-31	Admin. Request	House H.R. 3354	S. Cmte. S. 1603	P.L. 115-141		
Subtotal, CHIMPS	**-785.0**	**-831.0**	**-744.0**	**-1,032.0**	**-890.0**	**-941.0**	**-1.0**	**+743.0**	**-99.9%**
Rescissions (discretionary)	**-17.0**	**-34.0**	**-854.0**	**-899.7**	**-600.0**	**-800.0**	**-800.0**	**+54.0**	**-6.3%**
Other appropriations									
a. Disaster/emergency programs	116.0	273.0	234.8[k]	—	—	—	—	-234.8	-100.0%
b. Water and Waste Water	—	—	—	—	—	—	+500.0	+500.0	—
c. Broadband pilot	—	—	—	—	—	—	+600.0	+600.0	—
d. Opioid Enforcement and Surveillance	—	—	—	—	—	—	+94.0	+94.0	—
e. Food for Peace	—	250.0	134.0	—	—	—	+116.0	-18.0	-13.4%
f. Other appropriations	6.6	33.1	103.4	—	6.5	35.1	68.1	-35.3	-34.1%
Subtotal, Other appropriations	**122.6**	**556.1**	**472.2**	**—**	**6.5**	**35.1**	**1,378.1**	**+905.9**	**+191.9%**
Total, General Provisions	**-679.4**	**-308.9**	**-1,125.8**	**-1,931.7**	**-1,483.5**	**-1,705.9**	**577.1**	**+1,702.9**	**-151.3%**
Scorekeeping Adjustments[l]									
Disaster declaration in this bill	-116.0	-130.0	-206.1[k]	—	—	—	0.0	+206.1	-100.0%
Other scorekeeping adjustments	-398.0	-332.0	-524.0	-485.0	-481.0	-482.0	-481.0	+43.0	-8.2%
Subtotal, Scorekeeping adjustments	**-514.0**	**-462.0**	**-730.1**	**-485.0**	**-481.0**	**-482.0**	**-481.0**	**+249.1**	**-34.1%**
Totals									
Discretionary: Senate basis w/o CFTC	20,575.0	[21,500.0]	20,877.0	15,571.6	[19,748.9]	20,525.1	[23,002.0]	+2,125.0	+10.2%
Discretionary: House basis w/ CFTC	[20,825.0]	21,750.0	[21,127.0]	15,821.6	19,996.9	[20,775.1]	23,251.0	+2,124.0	+10.1%
Mandatory (M)	*126,756.5*	*118,990.7*	*132,506.9*	*124,918.3*	*124,915.4*	*124,915.4*	*122,752.0*	-9,754.9	-7.4%
Total: Senate basis w/o CFTC	*147,331.5*	*140,490.7*	*153,383.9*	*140,489.9*	*144,664.3*	*145,440.5*	*145,754.0*	-7,629.9	-5.0%
Total: House basis w/ CFTC	*147,581.5*	*140,740.7*	*153,633.9*	*140,739.9*	*144,912.3*	*145,690.5*	*146,003.0*	-7,630.9	-5.0%

the budget caps for discretionary spending in FY2018 and FY2019, provided disaster assistance for agriculture, amended several farm bill provisions, and extended sequestration effects.

Supplemental Appropriations and the Bipartisan Budget Act

On February 9, 2018, Congress passed the Bipartisan Budget Act of 2018 (BBA; P.L. 115-123), which broadly authorized supplemental appropriations, including for crop and livestock losses from 2017 hurricanes and wildfires (Division B, Subdivision 1, Title I). The act also included a six-week CR through March 23, 2018 (Division B, Subdivision 3). Importantly for the anticipated farm bill reauthorization, the BBA revised several agriculture programs, which had long-term policy implications because it changed farm bill statutes (Division B, Section 20101; and Division F) and added mandatory spending authority. Perhaps most importantly for completing the FY2018 appropriation, it raised the discretionary spending caps (Division C, Title I) that are in statute from the Budget Control Act of 2011 (BCA; P.L. 112-25).

Supplemental Appropriation

In supplemental appropriations, the BBA added $3.6 billion of disaster assistance in FY2018. Specifically, it provided $2.36 billion of block grants to the states for losses from 2017 hurricanes and wildfires. It added $941 million for conservation and watershed recovery, $165 million for rural water and wastewater recovery, and $89 million for disasters in six other USDA accounts.

Changes to the Farm Bill

For the expected successor to the current farm bill, the BBA added $1.4 billion of mandatory funding to the 10-year baseline. Specifically, it added $1.1 billion for dairy programs, $240 million for permanent disaster assistance programs, and a $62 million net addition for cotton. The cotton addition is nearly $3 billion from adding seed cotton to the farm commodity programs, offset by about $2.9 billion in reductions from reallocating base acres and crop insurance. A version of these changes was in the Senate appropriations markup, Section 728 of S. 1603.

The BBA offsets some of these additions by extending sequestration on mandatory accounts under the BCA (*Appendix A*) for two more years, for FY2026 and FY2027, at an estimated future effect on agriculture accounts of $2.6 billion.

Discretionary Budget Caps

Raising the budget caps for overall discretionary spending—that were set by the BCA— facilitated the development of a final full-year appropriation for FY2018. A majority in Congress desired greater spending for at least some of the appropriations subcommittees. For FY2018, the BBA raised the nondefense discretionary cap by $63 billion from $516 billion to $579 billion and for defense by $80 billion from $549 billion to $629 billion. For FY2019, it raised the nondefense discretionary cap by $68 billion and the defense cap by $85 billion.[25]

Consolidated Appropriations Act

On March 23, 2018, Congress passed the FY2018 Consolidated Appropriations Act (P.L. 115- 141).[26] The discretionary total of the Agriculture portion (Division A) is $23.3 billion, an increase of $2.1 billion above the enacted amount in FY2017 (+10%, on a comparable basis that includes CFTC; *Table 2*). The discretionary total is higher than in either the House-passed or Senate-reported bills, as became possible under the BBA, and is nearly as high as the peak in Agriculture appropriations in FY2010 (*Figure 3*).

---

[25] CRS Insight IN10861, Discretionary Spending Levels Under the Bipartisan Budget Act of 2018.

[26] The text of the Agriculture appropriation and explanatory material is published as Division A in the "House Committee Print, 115th Congress, Book 1—Consolidated Appropriations Act, 2018," April 27, 2018, https://www. gpo.gov/fdsys/pkg/CPRT-115HPRT29456/pdf/CPRT-115HPRT29456.pdf.

The appropriations also carry nearly $123 billion of mandatory spending, though that total is largely determined in separate authorizing laws. Thus, the overall total of the enacted FY2018 Agriculture appropriation is about $146 billion (*Figure 2*).

Compared to FY2017, the enacted appropriation increases spending primarily through an extra $1.38 billion in General Provisions, for programs that receive funding elsewhere in the appropriation. Amounts in General Provisions may not be as likely to become part of the annual base for those programs. For example, in addition to amounts for the Rural Utilities Service elsewhere in the act, the General Provisions provide an additional $500 million for rural water programs, and $600 million for expanding rural broadband. The General Provisions also provide an extra $116 million for Food for Peace foreign food aid, and $94 million for opioid enforcement and surveillance at FDA (*Table 3*).

In the regular portion of the appropriations for agencies, the Consolidated Appropriations Act increases agricultural research spending by $138 million over four agencies (including increasing the Agriculture and Food Research Initiative by $25 million to $400 million, and increasing buildings and facilities funding by $41 million for the Agricultural Research Service). It also increases the base amount for Food for Peace by $134 million to $1.6 billion.

Unlike the practice from more than the past decade, the FY2018 Agriculture appropriation does not impose as many changes to mandatory program spending (CHIMPS), such as to the Environmental Quality Incentives Program (EQIP), Fresh Fruit and Vegetable program, or rescissions to the Rural Development cushion of credit account or to Section 32. The absence of these usual CHIMPS in the FY2018 Agriculture appropriation costs about $740 million against the discretionary limit of the bill compared to FY2017.

In addition to setting budgetary amounts, the act is also a vehicle for policy-related provisions that direct how the executive branch should carry out the appropriation. These provisions may have the force of law if they are included in the text of the appropriation (usually in the General Provisions) but their effect is generally limited to the current fiscal year.

Table 3. Agriculture and Related Agencies Appropriations, by Agency, FY2015-FY2018 (budget authority in millions of dollars)

	FY2015	FY2016	FY2017	FY2018					
Agency or Major Program	P.L. 113-235	P.L. 114-113	P.L. 115-31	Admin. Request	House H.R. 3354	S. Cmte. S. 1603	P.L. 115-141	Change FY2017 to FY2018 Enacted	
Title I. Agricultural Programs									
Departmental Administration	364.5	373.2	403.9	376.9	280.5	398.2	396.0	-8.0	-2.0%
Research, Education and Economics									
Agricultural Research Service	1,177.6	1,355.9	1,269.8	993.1	1,194.1	1,182.4	1,343.4	+73.5	+5.8%
National Institute of Food and Agriculture	1,289.5	1,326.5	1,362.9	1,252.8	1,343.8	1,373.2	1,407.8	+44.9	+3.3%
National Agricultural Statistics Service	172.4	168.4	171.2	185.7	178.2	191.7	191.7	+20.5	+12.0%
Economic Research Service	85.4	85.4	86.8	76.7	76.8	86.8	86.8	+0.0	+0.0%
Under Secretary	0.9	0.9	0.9	0.9	0.8	0.9	0.8	-0.1	-10.4%
Marketing and Regulatory Programs									
Animal and Plant Health Inspection Service	874.5	897.6	949.4	812.9	907.8	956.4	985.1	+35.7	+3.8%
Agricultural Marketing Service[a]	82.4	82.5	86.2	78.6	76.7	90.2	156.8	+27.2[d]	+21.0%[d]
Section 32 (M)	1,284.0	1,303.0	1,322.0	1,344.0	1,344.0	1,344.0	1,344.0	+22.0	+1.7%
Grain Inspection, Packers, Stockyards Admin.[a]	43.0	43.1	43.5	43.0	42.9	43.5	moved to Ag Marketing Service[a]		
Under Secretary	0.9	0.9	0.9	0.9	0.8	0.9	0.9	+0.0	+0.0%
Food Safety									
Food Safety and Inspection Service	1,016.5	1,014.9	1,032.1	1,038.1	1,038.1	1,038.1	1,056.8	+24.8	+2.4%
Under Secretary	0.8	0.8	0.8	0.8	0.8	0.8	0.8	-0.0	-2.3%
Farm and Commodity Programs[a]									
Farm Service Agency[b]	1,603.3	1,595.1	1,624.0	moved to Title II: Farm Production and Conservation[a]					
FSA Farm Loans: Loan Authority[c]	*6,402.1*	*6,402.1*	*8,002.6*	moved to Title II: Farm Production and Conservation[a]					

Agency or Major Program	FY2015 P.L. 113-235	FY2016 P.L. 114-113	FY2017 P.L. 115-31	FY2018 Admin. Request	FY2018 House H.R. 3354	FY2018 S. Cmte. S. 1603	FY2018 P.L. 115-141	Change FY2017 to FY2018 Enacted	
Risk Management Agency Salaries and Exp.	74.8	74.8	74.8	moved to Title II: Farm Production and Conservation[a]					
Federal Crop Insurance Corporation (M)	8,930.5	7,858.0	8,667.0	moved to Title II: Farm Production and Conservation[a]					
Commodity Credit Corporation (M)	13,444.7	6,871.1	21,290.7	moved to Title II: Farm Production and Conservation[a]					
Under Secretary	0.9	0.9	0.9	0.9	—	—	—	-0.9	-100.0%
Subtotal									
Discretionary	**6,786.9**	**7,020.3**	**7,107.7**	**4,861.2**	**5,141.2**	**5,363.1**	**5,626.8**	**+219.0[d]**	**+4.0%[d]**
Mandatory (M; Section 32 in FY2018)	*23,659.7*	*16,032.6*	*31,280.2*	*1,344.0*	1,344.0	*1,344.0*	*1,344.0*	*+22.0[d]*	*+1.7%[d]*
Subtotal	*30,446.6*	*23,052.9*	*38,387.9*	*6,205.2*	6,485.2	*6,707.1*	*6,970.8*	*+240.5[d]*	*+3.6%[d]*
Title II. Farm Production and Conservation[a]									
Business Center	—	—	—	—	—	—	1.0	+1.0	—
Farm Service Agency[b]	moved from Title I[a]			1,508.8	1,555.4	1,627.0	1,625.2	+1.2	+0.1%
FSA Farm Loans: Loan Authority[c]	moved from Title I[a]			*6,953.9*	*7,163.9*	*8,002.6*	*8,005.6*	*+3.0*	*+0.0%*
Risk Management Agency Salaries and Exp.	moved from Title I[a]			55.0	55.0	74.8	74.8	+0.0	+0.0%
Federal Crop Insurance Corporation (M)	moved from Title I[a]			8,245.0	8,245.0	8,245.0	8,913.0	+246.0	+2.8%
Commodity Credit Corporation (M)	moved from Title I[a]			17,483.0	17,483.0	17,483.0	14,284.8	-7,005.9	-32.9%
Conservation Operations	846.4	850.9	864.5	766.0	864.5	874.1	874.1	+9.6	+1.1%
Watershed and Flood Prevention	—	—	150.0	—	42.0	150.0	150.0	+0.0	+0.0%
Watershed Rehabilitation Program	12.0	12.0	12.0	—	10.0	0.0	10.0	-2.0	-16.7%
Under Secretary	0.9	0.9	0.9	0.9	0.9	0.9	0.9	+0.0	+0.0%
Subtotal									
Discretionary	**859.3**	**863.8**	**1,027.4**	**2,330.2**	**2,527.8**	**2,726.4**	**2,735.6**	**+9.8[d]**	**+0.4%[d]**
Mandatory (M; CCC, crop insurance in FY2018)	moved from Title I I[a]			*25,728.5*	*25,728.5*	*25,728.5*	*23,198.3*	*-6,759.9[d]*	*-22.6%[d]*
Subtotal	moved from Title I I[a]			*28,058.7*	*28,256.3*	*28,454.9*	*25,933.9*	*-6,750.0[d]*	*-20.7%[d]*

Table 3. (Continued)

Agency or Major Program	FY2015	FY2016	FY2017	FY2018				Change FY2017 to FY2018 Enacted	
	P.L. 113-235	P.L. 114-113	P.L. 115-31	Admin. Request	House H.R. 3354	S. Cmte. S. 1603	P.L. 115-141		
Title III. Rural Development									
Salaries and Expenses (including transfers)[e]	678.2	682.9	675.8	624.0	656.6	675.8	680.8	+5.0	+0.7%
Rural Housing Service	1,298.4	1,616.4	1,654.9	1,365.3	1,465.7	1,565.1	1,582.4	-72.5	-4.4%
RHS Loan Authority[c]	*27,421.5*	*27,496.8*	*28,083.4*	*27,260.0*	*27,965.7*	*28,483.4*	*28,390.1*	*+306.7*	*+1.1%*
Rural Business-Cooperative Service[f]	103.2	90.5	97.7	—	83.6	95.5	109.5	+11.8	+12.1%
RBCS Loan Authority[c]	*984.5*	*979.3*	*988.4*	—	*877.1*	*988.4*	*991.2*	*+2.8*	*+0.3%*
Rural Utilities Service	501.7	559.3	639.9	5.4	478.6	612.4	628.1[g]	-11.7	-1.8%
RUS Loan Authority[c]	*7,464.1*	*8,210.6*	*8,217.0*	*6,217.0*	*8,217.0*	*8,217.0*	*8,219.9*	*+2.8*	*+0.0%*
Rural Economic Infrastructure Grants	—	—	—	161.9	122.7	—	—	—	—
Under Secretary	0.9	0.9	0.9	0.9	—	0.9	—	-0.9	-100.0%
Subtotal, Discretionary	**2,582.4**	**2,950.0**	**3,069.2**	**2,157.5**	**2,807.1**	**2,949.8**	**3,000.9[g]**	-68.3	-2.2%
Subtotal, RD Loan Authority[c]	***35,870.1***	***36,686.7***	***37,288.9***	***33,477.0***	***37,059.8***	***37,688.9***	***37,601.2***	*+312.3*	*+0.8%*
Title IV. Domestic Food Programs									
Child Nutrition Programs (M)	21,300.2	22,149.7	22,794.0	24,256.3	24,280.9	24,296.5	24,254.1	+1,460.2	+6.4%
WIC Program	6,623.0	6,350.0	6,350.0	6,150.0	6,150.0	6,350.0	6,175.0	-175.0	-2.8%
SNAP, Food and Nutrition Act Programs (M)	81,837.6	80,849.4	78,480.7	73,612.5	73,610.0	73,612.5	74,013.5	-4,467.2	-5.7%
Commodity Assistance Programs	278.5	296.2	315.1	293.6	317.1	317.1	322.1	+7.0	+2.2%
Nutrition Programs Administration	150.8	150.8	170.7	148.5	148.5	153.8	153.8	-16.9	-9.9%
Under Secretary	0.8	0.8	0.8	0.8	0.8	0.8	0.8	-0.0	-1.7%
Subtotal									
Discretionary	**7,094.1**	**6,838.9**	**6,884.7**	**6,615.9**	**6,664.5**	**6,887.9**	**6,709.8**	**-174.9**	**-2.5%**
Mandatory (M)	*103,096.7*	*102,958.1*	*101,226.7*	*97,845.8*	*97,842.9*	*97,842.9*	*98,209.6*	*-3,017.0*	*-3.0%*
Subtotal	*110,190.9*	*109,797.0*	*108,111.3*	*104,461.7*	*104,507.4*	*104,730.8*	*104,919.4*	*-3,191.9*	*-3.0%*

Agency or Major Program	FY2015 P.L. 113-235	FY2016 P.L. 114-113	FY2017 P.L. 115-31	FY2018 Admin. Request	FY2018 House H.R. 3354	FY2018 S. Cmte. S. 1603	FY2018 P.L. 115-141	Change FY2017 to FY2018 Enacted	
Title V. Foreign Assistance									
Foreign Agricultural Service	181.4	191.6	196.6	188.2	195.3	197.5	199.7	+3.1	+1.6%
Food for Peace Title II, and admin. expenses	1,468.5	1,468.5[g]	1,466.1[g]	0.1	1,400.1	1,600.1	1,600.1[g]	+134.0	+9.1%
McGovern-Dole Food for Education	191.6	201.6	201.6	0.0	201.6	206.6	207.6	+6.0	+3.0%
CCC Export Loan Salaries	6.7	6.7	8.5	6.7	6.7	8.5	8.8	+0.3	+3.6%
Office of Codex Alimentarius	—	—	—	—	—	—	3.8	+3.8	—
Under Secretary	—	—	—	—	0.9	0.9	0.9	+0.9	—
Subtotal	**1,848.3**	**1,868.5**	**1,872.9**	**195.1**	**1,804.7**	**2,013.7**	**2,021.0**	**+148.1**	**+7.9%**
Title VI. Related Agencies									
Food and Drug Administration	2,597.3	2,729.6	2,771.2	1,828.5	2,768.1	2,772.2	2,811.9	+40.7	+1.5%
Commodity Futures Trading Commission[h]	[250.0]	250.0	[250.0]	250.0	248.0	[250.0][i]	249.0	-1.0	-0.4%
Subtotal	**2,597.3**	**2,979.6**	**[3,021.2]**	**2,078.5**	**3,016.1**	**[3,022.2]**	**3,060.9**	**+39.7**	**+1.3%**
Title VII. General Provisions									
Reductions in Mandatory Programs[j]									
a. Environmental Quality Incentives Program	-136.0	-209.0	-179.0	-209.0	—	-179.0	—	+179.0	-100.0%
b. Watershed Rehabilitation Program	-69.0	-68.0	-54.0	-55.0	-55.0	-55.0	—	+54.0	-100.0%
c. Fresh Fruit and Vegetable Program	-122.0	-125.0	-125.0	-125.0	-125.0	-125.0	—	+125.0	-100.0%
d. Biorefinery Assistance Program	-16.0	-19.0	-20.0	-175.0	-175.0	-36.0	—	+20.0	-100.0%
e. Biomass Crop Assistance Program	-2.0	-20.0	-20.0	-20.0	-21.0	-21.0	-21.0	-1.0	+5.0%
f. Cushion of Credit (Rural Development)	-179.0	-179.0	-132.0	-176.0	-196.0	-196.0	—	+132.0	-100.0%
g. Section 32	-121.0	-216.0	-231.0	-263.0	-263.0	-263.0	—	+231.0	-100.0%
h. Other CHIMPS and mandatory rescissions	-140.0	+5.0	+17.0	-9.0	-55.0	-66.0	+20.0	+3.0	+17.6%

Source: CRS, using appropriations text and report tables, and unpublished Congressional Budget Office (CBO) tables. Reflects H.R. 3354 as amended from H.R. 3268.

Notes: Amounts are nominal discretionary budget authority in millions of dollars unless labeled otherwise; (M) indicates that the account is mandatory authority (or primarily mandatory authority). Excludes amounts in supplemental appropriations acts. Bracketed amounts are not in the official totals due to differing House-Senate jurisdiction for CFTC but are shown for comparison.

[a] Row headings reflect recent USDA reorganization. The Farm Service Agency and Risk Management Agency were moved from Title I to Title II, as was the Commodity Credit Corporation and Federal Crop Insurance Corporation in mandatory spending. Grain Inspection, Packers, and Stockyards Administration was moved into the Agricultural Marketing Service.

[b] Includes regular FSA salaries and expenses, plus transfers for farm loan program salaries and administrative expenses. Also includes farm loan program loan subsidy, State Mediation Grants, Dairy Indemnity Program (mandatory funding), and Grassroots Source Water Protection Program. Does not include appropriations to the Foreign Agricultural Service for export loans and P.L. 480 administration that are transferred to FSA.

[c] Loan authority is the amount of loans that can be made or guaranteed with a loan subsidy. This amount is not added in the budget authority subtotals or totals.

[d] Differences were computed using amounts for FY2017 that were adjusted to reflect the placement of agencies under the reorganization.

[e] Includes Rural Development salaries and expenses and transfers from the three rural development agencies for salaries and expenses. Amounts for the agencies thus reflect program funds for loans and grants.

[f] Amounts for the Rural Business-Cooperative Service (RBCS) are before the rescission in the Cushion of Credit account, unlike in Appropriations Committee tables. The rescission is included with the CHIMPS as classified by CBO, which allows the RBCS subtotal to remain positive.

[g] Excludes a portion of the other appropriations that are provided separately in General Provisions.

[h] Jurisdiction for CFTC is in the House Agriculture Appropriations Subcommittee and the Senate Financial Services Appropriations Subcommittee. After FY2008, CFTC is carried in enacted Agriculture appropriations in even-numbered fiscal years in House Agriculture markup but not in Senate Agriculture markup. Bracketed amounts are not in the official totals due to differing House-Senate jurisdiction for CFTC but are shown for comparison.

[I] Senate Financial Services and General Government Appropriations Subcommittee, chairman's draft, in lieu of subcommittee markup, November 20, 2017.

[j] Includes reductions (limitations and rescissions) to mandatory programs that may also be known as CHIMPS.

[k] Includes $206 million appropriated for the Emergency Conservation Program (ECP) and Emergency Watershed Program (EWP) in the FY2017 second CR (P.L. 114- 254, Section 185) that were offset as emergency spending. Another $29 million for ECP was included in the final appropriation (Section 753).

[l] "Scorekeeping adjustments" are not necessarily appropriated items and may not be shown in appropriations committee tables but are part of the official CBO score (accounting) of the bill. They predominantly include "negative subsidies" in loan program accounts and adjustments for disaster designations in the bill.

The explanatory statement that accompanies the appropriation—and the House and Senate report language that accompanies the committee-reported bills—may also provide policy instructions. These documents do not have the force of law but often explain congressional intent and are expected to be followed by the agencies. Indeed, the committee reports and explanatory statement may need to be read together to capture all of the congressional intent for the fiscal year.[27]

Table 3 presents the amounts in the FY2018 Consolidated Appropriations Act by agency and many programs, compared to the House-passed, Senate-reported, and Administration proposals. It also compares the FY2018 appropriation to three prior years, FY2015-FY2017.

APPENDIX A. BUDGET SEQUESTRATION

Sequestration is a process of automatic, largely across-the-board reductions that permanently cancel mandatory and/or discretionary budget authority.[28] Sequestration is triggered as a budget enforcement mechanism when federal spending would exceed statutory budget goals.[29] Sequestration is currently authorized in the BCA (P.L. 112-25) for discretionary spending through FY2021 and for mandatory spending through FY2027, as amended by subsequent acts and explained below.

Besides FY2013—when the timing of appropriations and the first year of sequestration resulted in triggering sequestration on discretionary spending—Bipartisan Budget Acts in 2013, 2015, and 2018 (P.L. 113-67, P.L. 114-74, and P.L. 115-123, respectively) have avoided sequestration on discretionary spending. These acts raised the discretionary budget caps that were placed in statute by the BCA and allowed Congress to enact larger appropriations than were allowed under the BCA. Sequestration, however,

---

[27] "House Committee Print, 115th Congress, Book 1—Consolidated Appropriations Act, 2018," at p. 63.

[28] CRS Report R43411, *The Budget Control Act of 2011: Legislative Changes to the Law and Their Budgetary Effects.*

[29] CRS Report R42972, *Sequestration as a Budget Enforcement Process: Frequently Asked Questions.*

continues to apply to certain accounts of mandatory spending and is not avoided by the Bipartisan Budget Acts (*Table A-1*).

The original FY2021 sunset on the sequestration of *mandatory* accounts has been extended four times to pay for avoiding sequestration of *discretionary* spending in the near term or as a general budgetary offset for other bills:

1. Congress extended the duration of mandatory sequestration by two years (until FY2023) as an offset in the Bipartisan Budget Act of 2013.[30]
2. Congress extended it by another year (until FY2024) to maintain retirement benefits for certain military personnel (P.L. 113-82).
3. Congress extended sequestration on nonexempt mandatory accounts another year (until FY2025) as an offset in the Bipartisan Budget Act of 2015.[31]
4. Congress extended sequestration on nonexempt mandatory accounts by another two years (until FY2027) as an offset in the Bipartisan Budget Act of 2018 (P.L. 115-123 Division C, Section 30101(c)).[32]

Some farm bill mandatory programs are exempt from sequestration. The nutrition programs and the Conservation Reserve Program are statutorily exempt,[33] and some prior legal obligations in crop insurance and the farm commodity programs may be exempt[34] as determined by OMB.[35]

---

[30] CBO, *Bipartisan Budget Act of 2013*, December 11, 2013, https://www.cbo.gov/publication/44964.

[31] CBO, *Bipartisan Budget Act of 2015*, October 28, 2015, https://www.cbo.gov/publication/50938. In addition to extending the duration of sequestration, Congress used crop insurance as a budgetary offset in the Bipartisan Budget Act of 2015. The effect was temporary, however, and the crop insurance reduction was restored. For more background, see the section on crop insurance and the Standard Reinsurance Agreement in CRS Report R44240, *Agriculture and Related Agencies: FY2016 Appropriations*.

[32] CBO, *Bipartisan Budget Act of 2018*, February 8, 2018, https://www.cbo.gov/publication/53556.

[33] 2 U.S.C. 905 (h), and 2 U.S.C. 905 (g)(1)(A).

[34] 2 U.S.C. 905 (g)(2), and 2 U.S.C. 906 (j).

Generally speaking, the experience since FY2013 is that OMB has ruled that most of crop insurance is exempt from sequestration, while the farm commodity programs, disaster assistance, and most conservation programs have been subject to it.[36]

For example, under the 2014 farm bill, the first farm commodity program payments began to be paid in October 2015, and USDA indicated that they would be subject to the 6.8% reduction that was applicable to FY2016.[37]

Table A-1. Sequestration from Discretionary and Mandatory Agriculture Appropriations (sequestered budget authority in millions of dollars)

	Discretionary Accounts		Mandatory Accounts	
Fiscal year	Rate	Amount	Rate	Amount
2013	5.0%	1,153	5.1%	713
2014	—	—	7.2%	1,052
2015	—	—	7.3%	1,153
2016	—	—	6.8%	1,819
2017	—	—	6.9%	1,686
2018	—	—	6.6%	1,316

Source: CRS, compiled from OMB, Reports to the Congress on the Joint Committee Reductions, various fiscal years. Available for FY2018 at https://www.whitehouse. gov/omb/legislative/omb-reports and FY2013-FY2017 at https://obamawhite house.archives.gov/omb/legislative_ reports/sequestration.

Notes: Sequestration rates listed here are for nonexempt, nondefense accounts. Amount totals were computed by CRS, as compiled in Table A-2.

Thus, sequestration on nonexempt *mandatory* accounts continues in FY2018. Nonexempt mandatory spending in agriculture accounts are reduced by a 6.6% sequestration rate and thus are paid at 93.4% of what they would have otherwise provided. This results in a reduction of about $1.3 billion less than what would have been authorized from mandatory agriculture accounts in FY2018.

---

[35] Some administrative expenses may be subject to sequestration, and therefore programs that are otherwise exempt may have a relatively small sequesterable amount compared to their total budget authority.

[36] CRS Report R42050, *Budget "Sequestration" and Selected Program Exemptions and Special Rules.*

[37] Southwest Farm Press, "Vilsack Announces 6.8% ARC/PLC Cuts Forthcoming," October 8, 2015, http://southwest farmpress.com/government/vilsack-announces-68-arcplc-cuts-forthcoming-2014-2016-payments-farmers.

Table A-1 shows the rates of sequestration that have been announced so far and the total amounts of budget authority that have been cancelled from accounts in the Agriculture appropriations bill. *Table A-2* provides additional detail at the account level for sequestration on mandatory accounts within the jurisdiction of Agriculture appropriations.

Table A-2. Sequestration of Mandatory Accounts in Agriculture Appropriations (sequestered budget authority in millions of dollars)

	FY2013	FY2014	FY2015	FY2016	FY2017	FY2018
Sequestration rate on nonexempt, nondefense mandatory accounts	*5.1%*	*7.2%*	*7.3%*	*6.8%*	*6.9%*	*6.6%*
U.S. Department of Agriculture						
Office of the Secretary	—	—	0.9	0.9	0.9	0.9
Office of Chief Economist	—	—	0.1	0.1	0.1	0.1
Agricultural Research Service	0.1	0.1	0.1	0.1	0.1	0.1
National Institute of Food and Agriculture	—	—	—	9.9	10.0	—
Extension	0.3	0.4	1.8	—	—	3.3
Biomass Research and Development	—	—	0.2	0.2	0.2	—
Integrated Activities	—	—	7.3	—	—	6.6
Animal and Plant Health Inspection Service						
Salaries appropriation	13.6	18.8	21.5	20.1	19.5	18.8
Miscellaneous Trust Funds	0.1	0.1	0.1	0.1	0.1	0.1
Food Safety Inspection Service	0.1	0.1	0.1	0.1	0.1	0.1
Grain Inspection Packers and Stockyards Administration	2.1	3.0	3.0	3.1	3.1	2.9
Agricultural Marketing Service						
Section 32	40.4	79.7	81.9	77.3	79.6	78.1
Milk Market Orders Assessment Fund	2.9	4.2	4.2	4.0	4.1	4.0
Perishable Ag Commodities Act	0.6	0.8	0.8	0.8	0.8	0.7
Expenses and refunds	0.4	0.9	0.9	1.3	1.3	0.4
Payments to States and Possessions	—	—	5.3	5.0	5.0	5.6
Marketing Services	—	—	2.2	2.0	2.1	2.0
Risk Management Agency	—	—	—	—	0.6	—
Federal Crop Insurance Corporation	3.0	4.2	5.9	3.5	3.9	3.7
Farm Service Agency						
Commodity Credit Corporation Fund	329.5	573.7	710.8	1,388.6	1,238.6	904.1
Agricultural Credit Insurance Corporation	—	—	0.1	0.1	0.1	0.1

Table A-2. (Continued)

	FY2013	FY2014	FY2015	FY2016	FY2017	FY2018
Commodity Credit Corporation Export Loans	—	—	—	0.4	0.4	0.3
Pima Cotton Trust Fund	—	—	—	1.1	1.1	1.1
Wool Apparel Manufacturers Trust Fund	—	—	—	2.0	2.1	2.0
Agricultural Disaster Relief Fund	70.0	—	—	—	—	—
Tobacco Trust Fund	49.0	69.1	—	—	—	—
Farm Security and Rural Investment Programs	171.2	263.1	269.9	265.7	281.5	254.9
Watershed Rehabilitation Program Rural Development, Rural Business Cooperative Service	—	11.9	11.2	4.7	4.7	4.4
Rural Energy for America Program	1.1	3.0	3.7	3.4	3.5	3.3
Rural Microenterprise Investment Program	—	0.2	0.2	0.2	0.2	0.2
Energy Assistance Payments	3.3	3.2	1.1	1.0	1.0	1.0
Biorefinery Assistance Program	—	—	3.7	3.4	1.4	1.3
Rural Economic Development Grants	—	—	—	1.6	—	—
Foreign Agricultural Service	0.1	0.1	0.1	0.1	0.1	0.1
Food and Nutrition Servicea						
SNAP	4.7	8.0	8.4	9.8	10.6	9.6
Child Nutrition Programs	2.5	4.2	4.2	3.9	4.3	4.0
Commodity Assistance Program	1.1	1.5	1.5	1.4	1.4	1.4
WIC	0.1	0.1	0.1	0.1	0.1	0.1
Related Agencies						
Food and Drug Administration						
Revolving Fund for Certification	0.4	0.6	0.6	0.6	0.6	0.6
User Fees	16.3	—	—	—	—	—
Farm Credit System Insurance Corporation	0.2	0.3	0.3	0.3	0.3	0.3
Commodity Futures Trading Commission	0.7	0.9	1.0	2.2	2.2	0.1
Total	713.3	1,051.9	1,153.0	1,818.9	1,685.6	1,315.9

Source: CRS, compiled from OMB, Reports to the Congress on the Joint Committee Reductions, various fiscal years. Available for FY2018 at https://www.whitehouse. gov/omb/legislative/omb-reports and FY2013-FY2017 at https://obamawhite house.archives.gov/omb/legislative_reports/sequestration.

Notes: Sequestration rates are for nonexempt, nondefense accounts. The sequesterable budget authority for each row may be computed by dividing the amount of sequestration by the sequestration rate. Column totals were computed by CRS.

[a] Benefits from the nutrition programs are generally exempt from sequestration by statute, but some administrative expenses in these programs may be subject to sequestration, and therefore a relatively small portion of the total budget authority may be sequesterable.

Table B-1. (Continued)

	House Action			Senate Action			Final Appropriation			
Fiscal Year	Subcmte.	Cmte.	Floor	Subcmte.	Cmte.	Floor	Enacted[a]		Public Law	CRS Report
2014	6/5/2013	6/13/2013	—	6/18/2013	6/20/2013	—	1/17/2014	O	P.L. 113-76	R43110
2015	5/20/2014	5/29/2014	—	5/20/2014	5/22/2014	—	12/16/2014	O	P.L. 113-235	R43669
2016	6/18/2015	7/8/2015	—	7/14/2015	7/16/2015	—	12/18/2015	O	P.L. 114-113	R44240
2017	4/13/2016	4/19/2016	—	5/17/2016	5/19/2016	—	5/5/2017	O	P.L. 115-31	R44588
2018	6/28/2017	7/12/2017	9/14/2017	7/18/2017	7/20/2017	—	3/23/2018	O	P.L. 115-141	R45128

Source: CRS.

[a] E = Enacted as standalone appropriation (seven times over 22 years); O = Omnibus appropriation (13 times); Y = Year-long CR (two times).

[b] A procedure that permits a Senate subcommittee to transmit a bill to its full committee without a formal markup session. See CRS Report RS22952, Proxy Voting and Polling in Senate Committee.

In: Agricultural Policy …
Editor: Albertus Brinkhuis

ISBN: 978-1-53615-215-9

Chapter 4

FY2018 and FY2019 Appropriations for Agricultural Conservation*

Megan Stubbs

Summary

The Agriculture appropriations bill funds the U.S. Department of Agriculture (USDA) except for the Forest Service. The FY2018 Consolidated Appropriations Act (P.L. 115-141, Division A), and both of the FY2019 agriculture bills reported by the House and Senate Appropriations Committees (H.R. 5961, S. 2976) include funding for conservation programs and activities at USDA.

Congress passed the FY2018 Consolidated Appropriations Act on March 23, 2018, which included agriculture appropriations under Division A. For FY2019, the House and Senate Appropriations Committees reported agriculture bills in May 2018. The Senate amended and passed its version as Division C of a four-bill minibus on August 1, 2018 (H.R. 6147). In the absence of a final appropriation, Congress enacted a continuing resolution through December 7, 2018 (P.L. 115-245, Division C).

* This is an edited, reformatted and augmented version of Congressional Research Service Publication No. R45406, Updated November 16, 2018.

Agricultural conservation programs include both mandatory and discretionary spending. Most conservation program funding is mandatory and is authorized in omnibus farm bills. Other conservation programs—mostly technical assistance—are discretionary and are funded through annual appropriations.

The largest discretionary program is the Conservation Operations (CO) account, which funds conservation planning and implementation assistance on private agricultural lands across the country. The enacted FY2018 appropriation provided $874 million for CO, an increase from the FY2017 enacted amount ($864 million). The FY2019 House-reported and Senate-passed bills would further increase funding for CO above FY2018 levels to $890 million and $879 million, respectively. Other discretionary spending is primarily for watershed programs. The largest—Watershed and Flood Prevention Operations (WFPO)—was funded at $150 million in FY2018. Both the House-reported and Senate-passed bills would fund WFPO at the $150 million level in FY2019.

Most mandatory conservation programs are authorized in omnibus farm bills and do not require an annual appropriation. However, Congress has reduced mandatory conservation programs through changes in mandatory program spending (CHIMPS) in the annual agricultural appropriations law every year since FY2003. The enacted FY2018 omnibus marks the first appropriation since FY2002 that does not include CHIMPS to mandatory conservation programs. For FY2019, both the House and Senate appropriation bills do not include reductions to mandatory conservation programs, because most programs' authorizations expired on September 30, 2018, making these programs ineligible for reduction.

While this is infrequent, the Agriculture appropriations bill may also serve as a vehicle for amendments to authorized programs that permanently alter or create programs. The FY2018 enacted appropriation included two such amendments—one to WFPO and one to farm bill conservation program reporting requirements. The WFPO amendment increased the size threshold required for congressional approval. Under the amended language, the Senate and House Agriculture Committees must approve WFPO projects that include an estimated federal contribution of more than $25 million for construction, an increase from the previous $5 million threshold. Additionally, the FY2018 appropriation exempted farm bill conservation programs from select federal reporting requirements, including obtaining a Data Universal Numbering System (DUNS) number and System for Award Management (SAM) registration.

Agriculture appropriations bills may also include policy-related provisions that direct how the executive branch should carry out the appropriation. The FY2018 enacted appropriation and both the FY2019 House-reported and Senate-passed bills include policy provisions for conservation programs that range from reports to Congress to suggested natural resource priorities.

INTRODUCTION

The Agriculture appropriations bill—formally known as the Agriculture, Rural Development, Food and Drug Administration, and Related Agencies Appropriations Act— funds all of the U.S. Department of Agriculture (USDA), excluding the U.S. Forest Service. Congress passed the FY2018 Consolidated Appropriations Act on March 23, 2018 (P.L. 115-141). The House and Senate Appropriations Committees have reported Agriculture appropriations bills for FY2019 (H.R. 5961, S. 2976). The Senate amended and passed its version as Division C of a four-bill minibus (H.R. 6147). The final appropriation for FY2019 is pending, and current funding is authorized under a continuing resolution (P.L. 115-245) through December 7, 2018. The enacted FY2018 appropriation and both FY2019 House-reported and Senate-passed bills include funding for conservation programs and related activities at USDA.

This chapter provides a brief overview of the conservation-related provisions in the FY2018 Agriculture appropriations act and FY2019 Agriculture appropriations bills.

CONSERVATION APPROPRIATIONS

USDA administers a number of agricultural conservation programs that assist private landowners with natural resource concerns. These include working land programs, land retirement and easement programs, watershed programs, technical assistance, and other programs. The two lead agricultural conservation agencies within USDA are the Natural Resources Conservation Service (NRCS), which provides technical assistance and administers most conservation programs, and the Farm

Service Agency (FSA), which administers the Conservation Reserve Program (CRP).[1]

Most conservation program funding is mandatory, obtained through the Commodity Credit Corporation (CCC) and authorized in omnibus farm bills (about $5.3 billion of CCC funds for conservation in FY2018).[2] Other conservation programs—mostly technical assistance—are discretionary spending and are funded through annual appropriations (about $1 billion annually).

For the first time since FY2002, the enacted FY2018 appropriation did not include reductions to mandatory conservation programs. It did, however, include legislative changes that affect farm bill programs and watershed programs. Similarly, both FY2019 House-reported and Senate-passed bills do not include reductions to mandatory conservation programs. However, pending farm bill reauthorization, most of these programs expired on September 30, 2018, and therefore are not eligible for reduction. The FY2018 appropriation included a slight increase from FY2017 levels for discretionary conservation programs. The FY2019 House-reported appropriations bill includes a slight increase, while the FY2019 Senate-passed appropriations bill includes a slight decrease (see *Table 1* and *Figure 1*).

Discretionary Conservation Programs

NRCS administers all discretionary conservation programs. The largest program and the account that funds most NRCS activities is Conservation Operations (CO). The CO account primarily funds Conservation Technical Assistance (CTA), which provides conservation planning and implementation assistance through field staff placed in almost all counties within the United States and territories. Other components of CO include

---

[1] For more information on individual conservation programs, see CRS Report R40763, *Agricultural Conservation: A Guide to Programs*.

[2] For more information on the CCC, see CRS Report R44606, *The Commodity Credit Corporation: In Brief*.

the Soil Surveys, Snow Survey and Water Supply Forecasting, and Plant Materials Centers.

The enacted FY2018 appropriation provided $874 million—more than the FY2017 enacted amount ($864 million). The FY2019 House-reported and Senate-passed bills would further increase funding for CO above FY2018 levels to $890 million and $879 million, respectively. The Trump Administration's FY2019 budget request ($699 million) is below the House-reported and Senate-passed bills for FY2019 due to a proposed consolidation of mandatory and discretionary accounts to pay for conservation technical assistance.[3] The proposal to consolidate funding has been made by USDA through multiple Administrations but never adopted by Congress (see text box below).

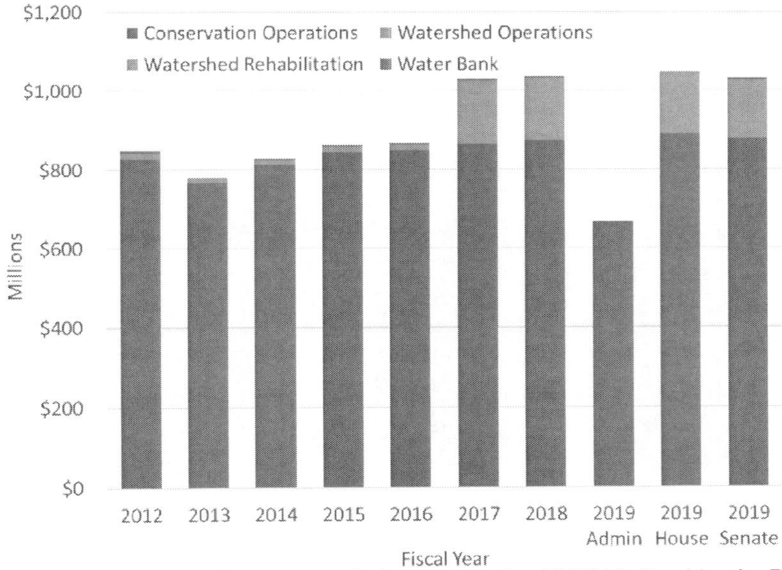

Source: CRS, from previous appropriations acts, the FY2019 President's Budget Request, H.R. 5961, and H.R. 6147.

Figure 1. Discretionary Appropriations for Agricultural Conservation Programs.

[3] Technical assistance is currently funded through both mandatory and discretionary sources. Because the amount of funding for technical assistance from mandatory funding sources is generally not reported, it is unknown whether the Administration's FY2019 proposal to consolidate funding from mandatory and discretionary sources would represent an increase or decrease in overall funding for conservation technical assistance.

Table 2. Selected Conservation Policy Provisions in the FY2018 and FY2019 Appropriations Bill Text

FY2018	**FY2019**	
Enacted, P.L. 115-141	House-reported, H.R. 5961	Senate-passed, H.R. 6147
Conservation Operation. Directs $5.6 million of CO to WFPO projects providing water to rural communities (Title II).	No comparable provision.	Same as FY2018 enacted (Title II).
Watershed Operations. Limits the application of the 250,000-acre limitation in WFPO to only activities where the primary purpose is flood prevention (Title II).	Similar to FY2018 enacted (Title II).	Similar to FY2018 enacted (Title II).
Directs $50 million of available funds to be allocated to projects that commence promptly, address select regional priorities, or are authorized under the Flood Control Act of 1944 (Title II).	Similar to FY2018 enacted(Title II).	Similar to FY2018 enacted(Title II).
Watershed Rehabilitation. Directs $5 million to states with high-hazard dams that have incurred fatal flooding events (Title II).	Similar to FY2018 enacted(Title II).	No comparable provision.
Agricultural Management Assistance (AMA).	Similar to FY2018	Similar to FY2018 enacted(§707).
Allows AMA funds to remain available until expended(§707).	enacted(§707).	
Conservation Reserve Program. Provides $1 million for a CRP bottomland hardwood tree pilot program (§743).	No comparable provision.	Similar to FY2018 enacted (§739).

(S.Rept. 115-131) that remain unchanged by this agreement, except as noted in this explanatory statement.... The House and Senate report language that is not changed by the explanatory statement is approved and indicates congressional intentions. The explanatory statement, while repeating some report language for emphasis, does not intend to negate the language referred to above unless expressly provided herein." U.S. Congress, House Committee on Appropriations, *Committee Print on H.R. 1625/Public Law 151- 141, Book 1—Consolidated Appropriations Act, 2018*, 115th Congress, 2nd session, 29-456 (Washington, DC: GPO, 2018), https://www.gpo.gov/fdsys/pkg/CPRT-115HPRT29456/pdf/CPRT-115HPRT29456.pdf.

Table 1. FY2017-FY2019 Agricultural Conservation Funding (budget authority in thousands of dollars)

Program	**FY2017**		**FY2018**				**FY2019**		
	P.L. 115-31	Admin. Request	House H.R. 3354	Senate S. 1603	P.L. 115-141	Admin. Request	House- Reported H.R. 5961	Senate-Passed H.R. 6147	Enacted
Conservation Operations	864,474	766,000	858,911	874,107	874,107	669,033	890,293	879,107	—
Conservation Technical Assistance	759,211	667,675	760,211	768,844	774,444	575,862	790,912	773,844	—
Soil Survey	80,802	79,696	80,000	80,802	80,802	74,438	80,500	80,802	—
Snow Survey	9,380	9,265	9,300	9,380	9,380	9,316	9,400	9,380	—
Plant Material Center	9,481	9,364	9,400	9,481	9,481	9,417	9,481	9,481	—
Watershed Projects (Watershed Operations)	5,600	0	0	5,600	5,600	0	0	5,600	—
Total Conservation Operations	864,474	766,000	858,911	874,107	847,107	669,033	890,293	879,107	—
Watershed Operations	150,000	0	40,000	150,000	150,000	0	150,000	150,000	—
Watershed Rehabilitation Program	12,000	0	10,000	0	10,000	0	10,000	0	—
Water Bank	4,000	0	0	4,000	4,000	0	0	4,000	—
Total NRCS Discretionary	1,030,474	766,000	908,911	1,028,107	1,038,107	669,033	1,050,293	1,033,107	—

Source: CRS, using appropriations text and report tables.

Notes: Amounts are nominal discretionary budget authority in thousands of dollars unless labeled otherwise. Excludes amounts in supplemental appropriations acts and proposed rescission language.

The FY2018 enacted appropriation and both FY2019 bills direct CO funding for a number of conservation programs (*Table 1*). Report language further directs funding to selected activities (*Table 2*). The enacted FY2018 appropriation also contains funding for watershed activities, including $150 million for Watershed and Flood Prevention Operations (WFPO)—a program that assists state and local organizations with planning and installing measures to prevent erosion, sedimentation, and flood damage.[4] This is the same level as appropriated in FY2017, which was the first appropriated funding for the WFPO program since FY2010. Beginning in FY2006, Administrations began requesting no funding for WFPO, citing program inflexibility and a backlog of congressionally earmarked projects. The Administration's FY2018 and FY2019 requests proposed no funding for the program. The FY2018 enacted appropriation and House-reported and Senate-passed FY2019 bills each include $150 million for WFPO.

Since FY2014, Congress has directed a portion of CO funds to select WFPO activities. Similar directive language ($5.6 million; see *Table 1*) is in the FY2018 appropriations, in addition to the $150 million made available for the program as a whole. The Senate-passed bill continues this directive language for watershed projects, while the House-reported bill does not.

The enacted FY2018 appropriation includes $10 million for the Watershed Rehabilitation program—a reduction from the FY2017 level of $12 million. The Watershed Rehabilitation program repairs aging dams previously built by USDA under WFPO. The Administration proposed no funding in FY2018 and FY2019. The FY2019 House-reported bill includes $10 million for Watershed Rehabilitation, while the FY2019 Senate-passed bill does not fund the program. The 2014 farm bill (P.L. 113-79) provided additional mandatory funding for the program ($250 million in FY2014) to remain available until expended.

[4] For additional information, see CRS Report RL30478, *Federally Supported Water Supply and Wastewater Treatment Programs*.

Funding Technical Assistance for Mandatory Conservation Programs

Most conservation technical assistance provided by NRCS is funded through the CTA program within CO. The funds are used to support voluntary conservation efforts at the local level through NRCS field staff. Technical assistance is also a part of the farm bill conservation programs, which are funded through a mandatory authorization. Most technical assistance activities within mandatory programs are in support of delivering some level of financial assistance as part of a contract or agreement. These activities could include providing designs, standards, and specifications needed to install scheduled conservation practices and activities. Generally, technical assistance prior to a producer entering into a contract for financial assistance is considered to be part of CTA. It is not until after a producer signs a contract for financial assistance that technical assistance is funded from the individual mandatory program rather than CTA. Once the financial assistance contract is complete, most mandatory program funds are no longer available to support ongoing assistance in maintaining the conservation plans, practices, and activities implemented under the financial assistance program.

Issues with the process of funding technical assistance for mandatory programs date back to the mid-1990s. Multiple Congresses and Administrations have proposed funding technical assistance in different ways than the current method. The Administration's FY2019 budget request includes a proposal to create a consolidated account that would include transferred funding from mandatory conservation programs as well as discretionary appropriations for the purposes of providing technical assistance to farm bill conservation programs. This is not a new concept: It was also included in the FY2018 (Trump) and FY2017-FY2014 (Obama) presidential budget requests.

Mandatory Conservation Programs

Mandatory conservation programs are generally authorized in omnibus farm bills and receive funding from the CCC and thus do not require an annual appropriation.[5] But Congress used annual agriculture appropriations acts to reduce mandatory conservation programs through changes in mandatory program spending (CHIMPS) every year from FY2003 to FY2017.[6] The enacted FY2018 omnibus marks the first appropriation since FY2002 that does not include CHIMPS to conservation programs. This allowed all mandatory conservation programs to utilize their full authorized level of funding in FY2018, minus sequestration.[7] Additionally, prior-year CHIMPS concerning programs that are authorized to remain available until expended (e.g., Watershed Rehabilitation) remained available for obligation in FY2018.

Nearly all mandatory conservation programs authorized in the 2014 farm bill (Agricultural Act of 2014; P.L. 113-79) expired on September 30, 2018.[8] One exception is the Environmental Quality Incentives Program (EQIP), whose authority was extended to FY2019 in the Bipartisan Budget Act of 2018 (BBA; P.L. 115-123).[9] Similar to FY2018, the FY2019 House-reported and Senate-passed bills do not include reductions to mandatory conservation programs. Most are not authorized in FY2019 and therefore are ineligible for reduction.

---

[5] For authorized funding levels for mandatory conservation programs, see CRS Report R40763, *Agricultural Conservation: A Guide to Programs*.

[6] For additional background, see CRS In Focus IF10041, *Reductions to Mandatory Agricultural Conservation Programs in Appropriations Law*.

[7] For additional discussion on sequestration, see Appendix A of CRS Report R45230, *Agriculture and Related Agencies: FY2019 Appropriations*.

[8] For additional information on the expiration of the 2014 farm bill, see CRS Report R45341, *Expiration of the 2014 Farm Bill*.

[9] EQIP was CHIMPed in the FY2017 appropriation (P.L. 115-31), and the CHIMP carried over into the first half of FY2018 under continuing resolutions. Because the Congressional Budget Office (CBO) uses the last year of authority to develop the funding baseline for farm bill reauthorization (i.e., FY2018), the reduced EQIP authority could have affected the overall farm bill baseline. With the BBA extension to FY2019 and the passage of the FY2018 appropriations without a CHIMP, the full EQIP authority was ultimately used to develop the CBO farm bill baseline, which is in turn used by authorizing committees to develop the next omnibus farm bill. For more information on the farm bill baseline and how it is developed, see CRS Report R44758, *Farm Bill Programs Without a Budget Baseline Beyond FY2018*.

Amendments to Conservation Programs

Generally, Congress employs two separate types of legislative measures—authorization and appropriations. Authorization acts establish, continue, or modify agencies or programs.

Appropriations acts generally provide discretionary funding for authorized agencies and programs.[10]

While this practice is infrequent and subject to various procedural rules and limitations, the Agriculture appropriations bill may serve as a vehicle for amendments to authorized programs that permanently alter or create programs. These amendments generally have the force of law by amending the *U.S. Code* or by creating a permanent authorization. This is different from policy-related provisions (discussed in the "Policy-Related Provisions" section), which generally direct how the executive branch should carry out the appropriation and whose effect is typically limited to the current fiscal year.[11]

Watershed and Flood Prevention Operations

The FY2018 enacted appropriation included statutory amendments to the WFPO program. Section 761 of P.L. 115-141 amended the Watershed Protection and Flood Prevention Act of 1954 (16 U.S.C. 1001 *et seq.*) by increasing the size thresholds required for congressional approval under the program. Under the amended language, approval by the Senate and House Agriculture Committees is required for individual projects that need an estimated federal contribution of more than $25 million for construction, an increase from the previous $5 million threshold. This amendment originated in the FY2018 Senate-reported bill (S. 1603, §754).

---

[10] For additional information, see CRS Report R42388, *The Congressional Appropriations Process: An Introduction.*

[11] For more information on the authorization and appropriations process, see CRS Report RS20371, *Overview of the Authorization-Appropriations Process.*

Conservation Program Requirements

The FY2018 appropriation also amended Title XII of the Food Security Act of 1985 (P.L. 99-198; often referred to as the "1985 farm bill") by adding a new section that exempts farm bill conservation programs from certain reporting requirements. Federal grant recipients must comply with government-wide financial management policies and reporting requirements when receiving federal grants and agreements. Many of these reporting requirements are not new for USDA programs and have been in place for a number of years.

Interested stakeholders raised concerns when a number of the farm bill conservation programs were designated as grants (rather than direct payments) under a 2010 regulation. This designation triggered the use of a Data Universal Numbering System (DUNS) number and System for Award Management (SAM) registration.[12] The DUNS number requirement and SAM registration did not affect individuals or entities that apply for conservation programs using a Social Security number. Rather, it applied only to those applying as an entity with a Taxpayer Identification Number or Employee Identification Number.

The amendment exempts producers and landowners who participate in farm bill conservation programs from the DUNS number and SAM registration requirement.[13] The amendment originated in the FY2018 Senate-reported bill (S. 1603, §740).

POLICY-RELATED PROVISIONS

In addition to setting budgetary amounts, the Agriculture appropriations bill may also include policy-related provisions that direct

---

[12] For additional information on federal grant reporting requirements, see CRS Report R44374, *Federal Grant Financial Reporting Requirements and Databases: Frequently Asked Questions*.

[13] NRCS, *DUNS and SAM Update: 2018 Omnibus Act Nixes Requirement for Farmers*, March 23, 2018, https://www.nrcs.usda.gov/wps/PA_NRCSConsumption/download?cid=nrcseprd14 00 413&ext=pdf.

how the executive branch should carry out an appropriation. These provisions may have the force of law if they are included in the text of an appropriations act, but their effect is generally limited to the current fiscal year (see *Table 2*). Unlike the aforementioned authorization amendments that may be included in appropriations, policy-related provisions generally do not amend the *U.S. Code* or have long-standing effects.

For example, the WFPO program has historically been called the "small watershed program," because no project may exceed 250,000 acres, and no structure may exceed more than 12,500 acre-feet of floodwater detention capacity or 25,000 acre-feet of total capacity. The FY2018 enacted appropriation also included a policy provision that waives the 250,000-acre project limit when the project's primary purpose is something other than flood prevention. This provision did not amend the WFPO authorization and therefore was effective only for the funds provided during the appropriation year.[14] Both the FY2019 House-reported and Senate-passed bills include similar language that would continue this waiver for funds provided in FY2019.

Table 2 compares some of the policy provisions that have been identified in the Farm Production and Conservation Programs (Title II) and General Provisions (Title VII) titles of the FY2018 and FY2019 Agriculture appropriations bills related to conservation. Many of these provisions were also included in past years' appropriations laws.

The explanatory statement that accompanies the final appropriation—and the House and Senate report language that accompanies the committee-reported bills—may also provide policy instructions. These documents do not have the force of law but often explain congressional intent, which the agencies are expected to follow (see *Table 3*). The committee reports and explanatory statement may need to be read together to capture all of the congressional intent for a given fiscal year.[15]

---

[14] The provision applies to the $150 million appropriated in FY2018 and any funds previously provided. Since WFPO funding is available until expended, it is possible that the waiver could carry forward into future fiscal years but only for funds made available in FY2018 and before.

[15] According to the FY2018 Explanatory Statement, "The explanatory statement is silent on provisions that were in both the House Report (H.Rept. 115-232) and Senate Report

U.S. food supply, including rising demand for products produced outside the United States, as well as other developments, including emerging microbial pathogens, natural toxins, and technological innovations in production and processing.

Although Congress authorized funds to be appropriated in FSMA, it did not provide the full funding needed for FDA to perform these expanded activities under the law, according to FDA.[9] Prior to enactment, the Congressional Budget Office (CBO) estimated that fully implementing FSMA could increase net federal spending subject to appropriation by about $1.4 billion over a five-year period (FY2011-FY2015).[10] This cost estimate covered activities at FDA and other federal agencies and did not include offsetting revenue from the collection of new user fees authorized under FSMA.

Following FSMA's enactment, FDA officials continued to emphasize the need for additional FDA funding.[11] Food industry groups also called for increased budget authority for FDA to fully implement FSMA, at levels requested by the Administration, in order to maintain consumer confidence.[12] State agriculture officials and representatives of the National Association of State Departments of Agriculture (NASDA) further pushed for full FSMA funding to support implementation by front-line state officials.[13] Public health and consumer safety groups, as well as victims of food-borne illnesses, have also continued to call for additional food safety funding.[14]

---

[9] See, for example, FDA, Building Domestic Capacity to Implement the FDA Food Safety Modernization Act (FSMA), May 2013.

[10] CBO, cost estimate, "S. 510, Food Safety Modernization Act, as Reported by the Senate Committee on Health, Education, Labor, and Pensions on December 18, 2009, Incorporating a Manager's Amendment Released on August 12, 2010," August 12, 2010. Estimated total costs would be covered by a combination of user fees and direct appropriations (budget authority).

[11] See, for example, testimony to the Senate Appropriations Committee, from Stephen Ostroff, Acting Commissioner, FDA Food and Drugs, September 16, 2015.

[12] See, for example, letter dated April 15, 2014, to House and Senate committee appropriators from 20 industry groups, including Costco Wholesale Corporation, Wal-Mart Stores, GMA, and the Produce Marketing Association.

[13] See letter dated September 21, 2015, to House and Senate committee appropriators from NASDA.

[14] See, for example, letters (dated October-November 2014) to House and Senate committee appropriators from the relatives of victims of eating contaminated foods and from several

Table 2. (Continued)

FY2018	FY2019	
Water Bank. Provides $4 million for the Water Bank program (§745).	No comparable provision.	Similar to FY2018 enacted (§741).
Hardwood Tree Pilot. Provides $600,000 for a pilot program for nonindustrial private forest lands in Gulf Coast states impacted by Hurricane Katrina in 2005(§767).	No comparable provision.	No comparable provision.
Enacted, P.L. 115-141	House-reported, H.R. 5961	Senate-passed, H.R. 6147
No comparable provision.	No comparable provision.	Ocelots. Requires a report to Congress on ocelot (wild cat) conservation through NRCS programs (§762).

Source: CRS.

Table 3. Selected Conservation Policy Provisions in the FY2018 and FY2019 Appropriations Report Language

FY2018	FY2019	
Enacted Report	House Committee Report, H.Rept. 115-706	Senate Committee Report, S.Rept. 115-259
Conservation Operations. Directs NRCS to provide flexibility to state conservation officers in determining human resource needs (E).	No comparable provision.	No comparable provision.
Agricultural Conservation Easement Program (ACEP). Encourages NRCS to work with state and local partners to address ecological needs (H). Concerned about delays and deed terms (S).	Similar to FY2018 enacted.	No comparable provision.
Cheat Grass Eradication. Encourages NRCS to assist with cheat grass eradication, control, and fuel reduction (H).	Similar to FY2018 enacted.	No comparable provision.
Community Colleges. Encourages NRCS to collaborate with community colleges on conservation technology (H).	Similar to FY2018 enacted.	No comparable provision.
Floodplain Buyouts. Encourages NRCS to consider the unintended consequences of floodplain buyouts (H).	Similar to FY2018 enacted.	No comparable provision.
Harmful Algal Blooms. Supports NRCS's	Similar to FY2018 enacted.	No comparable provision.

FY2018	FY2019	
Enacted Report	House Committee Report, H.Rept. 115-706	Senate Committee Report, S.Rept. 115-259
soil erosion prevention efforts (H).		
Herbicide Resistance. Encourages NRCS to ensure staff are aware of herbicide resistant weed challenges (H).	Similar to FY2018 enacted.	No comparable provision.
Irrigation. Encourages NRCS to expand irrigation infrastructure assistance (H).	No comparable provision.	No comparable provision.
Enacted Report	House Committee Report, H.Rept. 115-706	Senate Committee Report, S.Rept. 115-259
Locally Led Conservation. Encourages collaboration between NRCS and state, local, tribal, and partners (H).	Similar to FY2018 enacted.	No comparable provision.
Milkweed. Encourages NRCS to increase benefits for milkweed and monarch butterfly habitat (H).	Similar to FY2018 enacted.	No comparable provision.
Regional Conservation Partnership Program (RCPP). Encourages NRCS to consider organic producers' needs under RCPP (H). Concerned about technical assistance reimbursement (S).	Similar to FY2018 enacted.	No comparable provision.
Resource Conservation and Development Councils (RC&Ds). Encourages NRCS to continue working with RC&Ds (H & S).	Similar to FY2018 enacted.	No comparable provision.
Sage Grouse Initiative. Supports the initiative (H).	Similar to FY2018 enacted.	No comparable provision.
Continuous CRP. Encourages the enrollment of State Acres for Wildlife Enhancement practices under CRP (S).	No comparable provision.	Similar to FY2018 enacted.
Wetlands Mitigation. Encourages USDA to use a 1-to-1 acre ratio for wetlands mitigation requirements (S).	No comparable provision.	Similar to FY2018 enacted.
Agricultural Management Assistance (AMA) Program. Encourages USDA to establish pilot projects related to food safety (S).	No comparable provision.	Similar to FY2018 enacted.
Salton Sea. Encourages NRCS to work with the U.S. Army Corps of Engineers and the State of California to restore the Salton Sea (S).	No comparable provision.	No comparable provision.
Technical Assistance. Directs NRCS to record and report total technical assistance levels to Congress (S).	No comparable provision.	Similar to FY2018 enacted.

Table 3. (Continued)

FY2018	FY2019	
CRP Commodity Food Plots. Directs the CCC to amend CRP policies and practices to permit current and future participants to plant but not harvest agricultural commodity crops as wildlife food plots on up to 10% of the enrolled land (S).	No comparable provision.	Similar to FY2018 enacted.
No comparable provision.	National Marine Sanctuaries (NMS). Encourages NRCS to continue to collaborate with NMS.	No comparable provision.
No comparable provision.	Irrigation Water Use. Encourages USDA to develop new conservation and irrigation techniques to reduce agricultural water use.	No comparable provision.
No comparable provision.	No comparable provision.	Conservation Practices. Encourages NRCS to prioritize EQIP practices that score highly on the Conservation Practices Physical Effects matrix.
No comparable provision.	No comparable provision.	NRCS Staffing. Directs NRCS to provide staffing levels to Congress.
No comparable provision.	No comparable provision.	Program Duplication. Directs NRCS to report to Congress on program duplication identified in inspector general reports.

Source: CRS.

Note:

[a] Provisions listed in the enacted FY2018 Explanatory Statement are cited as (E) and may be found at U.S. Congress, House Committee on Appropriations, Committee Print on H.R. 1625/Public Law 151-141, Book 1—Consolidated Appropriations Act, 2018, 115th Congress, 2nd session, 29-456 (Washington, DC: GPO, 2018), https://www.gpo.gov/fdsys/pkg/CPRT-115HPRT29456/pdf/CPRT-115HPRT 29456.pdf. Provisions listed in the House report (H.Rept. 115-232) are cited as (H). Provisions listed in the Senate report (S.Rept. 115-131) are cited as (S). All provisions within this column appear under Title II of the cited report.

Table 3 compares some of the policy provisions that have been identified in the FY2018 and FY2019 Agriculture appropriations reported language related to conservation. The FY2018 enacted report language column includes references to the House (H) and Senate (S) report language, as well as the enacted (E) explanatory statement to better capture congressional intent for FY2018. Many of these provisions have been included in past years' appropriations laws. Some provisions in report language and bill text address conservation programs not authorized or funded within the annual appropriation (i.e., mandatory spending for farm-bill-authorized programs).

In: Agricultural Policy ...
Editor: Albertus Brinkhuis

ISBN: 978-1-53615-215-9

Chapter 5

FY2018 AND FY2019 AGRICULTURE APPROPRIATIONS: FEDERAL FOOD SAFETY ACTIVITIES*

Renée Johnson and Sahar Angadjivand

SUMMARY

The Agriculture appropriations bill—formally known as the Agriculture, Rural Development, Food and Drug Administration, and Related Agencies Appropriations Act—funds the Food and Drug Administration (FDA) and the U.S. Department of Agriculture (USDA), excluding the U.S. Forest Service. Congress enacted the FY2018 agriculture appropriation in March 2018 as part of the Consolidated Appropriations Act, 2018 (P.L. 115-141, Division A). Both the House and the Senate Appropriations Committees have reported Agriculture appropriations bills for FY2019 (H.R. 5961, S. 2976). The Senate amended and passed its version as Division C of a four-bill minibus (H.R. 6147).

* This is an edited, reformatted and augmented version of Congressional Research Service Publication No. R45413, Updated November 27, 2018.

Numerous federal, state, and local agencies share responsibilities for regulating the safety of the U.S. food supply. Federal responsibility for food safety rests primarily with FDA, an agency of the Department of Health and Human Services, and also the Food Safety and Inspection Service (FSIS), an agency of USDA. FDA is responsible for ensuring the safety of the majority of all domestic and imported food products—except for meat and poultry products, which are within USDA's jurisdiction to oversee meat, poultry, and processed egg products.

Appropriations for Federal Food Safety Activities, FY2009-FY2018

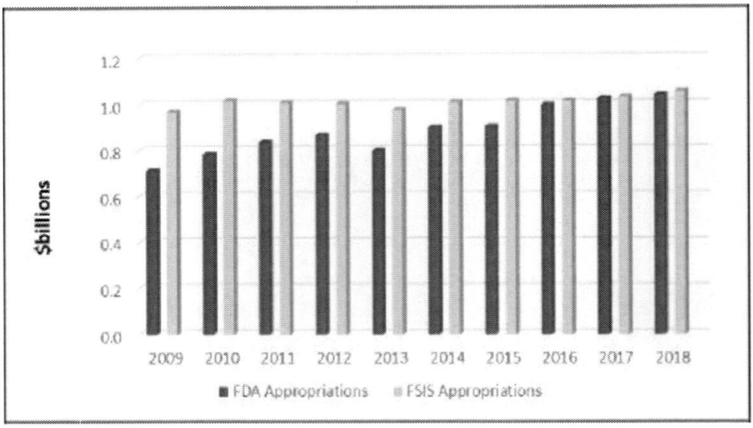

Combined appropriations covering food safety activities at both FDA and USDA totaled nearly $2.1 billion in FY2018. Congressional appropriations at both FDA and USDA are augmented by existing (currently authorized) user fees. FDA user fees authorized by the FDA Food Safety Modernization Act (FSMA, P.L. 111-353) have generated between $10 million and $18 million annually in recent years. At FSIS, user fees have generated between $180 million and $250 million per year.

At FDA, ongoing efforts to improve food safety include implementation of FSMA. FSMA was enacted by the 111th Congress and was the largest expansion of FDA's food safety authorities since the 1930s. Since FSMA became law in 2011, congressional appropriators have increased annual funding for the FDA Foods Program by $204.3 million—an increase of about 24% between FY2011 and FY2018—largely in an effort to support FDA's implementation of FSMA. The enacted FY2018 appropriation for FDA's Foods Program provided $1,041.6 million. Currently, FDA funding for its food safety oversight activities is roughly similar to those of FSIS (*Figure 1*). FDA's total budget for food safety programs and activities extends beyond the agency's Foods Program, encompassing other food and veterinary medicine programs at FDA.

Food-safety-related activities at FSIS include continuous inspections at federal meat and poultry plants. The enacted FY2018 appropriation provided $1,056.8 million to carry out this function. Compared to FY2011—the year FSMA was enacted— annual congressional appropriations for FSIS have increased by $48.3 million (+5%).

For FY2019, congressional appropriators would increase funding for federal food safety activities, whereas the Administration's budget proposal would reduce food safety funding below FY2018 levels at both FDA and FSIS.

INTRODUCTION

The Agriculture appropriations bill—formally known as the Agriculture, Rural Development, Food and Drug Administration, and Related Agencies Appropriations Act— funds the Food and Drug Administration (FDA) and the U.S. Department of Agriculture (USDA), excluding the U.S. Forest Service. This includes funding for FDA and USDA's Food Safety and Inspection Service (FSIS), the two primary federal agencies responsible for overseeing the safety of the nation's food supply.

In March 2018, Congress enacted the FY2018 agriculture appropriation as part of the Consolidated Appropriations Act, 2018 (P.L. 115-141, Division A). Both the House and the Senate Appropriations Committees have reported Agriculture appropriations bills for FY2019 (H.R. 5961, S. 2976). The Senate amended and passed its version as Division C of a four-bill minibus (H.R. 6147). The final appropriation for FY2019 is pending, and current funding is authorized under a continuing resolution (P.L. 115-245) through December 7, 2018. The enacted FY2018 appropriation and both the FY2019 House-reported bill and the Senate-passed bill include funding for food safety programs and related activities at FDA and USDA.

This chapter provides a brief overview of the FY2018 and FY2019 appropriations that address food safety activities at FDA and FSIS. It does not specifically address funding levels for other federal agencies or other USDA agencies that may play a role in ensuring the safety of the nation's food supply.[1]

Numerous federal, state, and local agencies share responsibilities for regulating the safety of the U.S. food supply.[2] Federal responsibility for food safety rests primarily with FDA and FSIS. FDA, an agency of the Department of Health and Human Services (HHS), is responsible for ensuring the safety of the majority of all domestic and imported food products (except for meat and poultry products).[3] FSIS, an agency at USDA, regulates most meat, poultry, and processed egg products.[4] Roughly speaking, FSIS regulates about 10%-20% of the U.S. food supply, while FDA is responsible for the remaining 80%-90%.[5]

Combined appropriations covering food safety activities at both FDA and USDA totaled nearly $2.1 billion in FY2018, which is roughly split between the two agencies (*Figure 1, Table 1*). This funding distribution reflects greater increases in congressional appropriations for FDA compared with FSIS since FY2011. The enacted FY2018 appropriation for FDA's food safety activities provided $1,041.6 million. For FSIS, the FY2018 appropriation provided $1,056.8 million.

---

[1] For example, USDA's Animal and Plant Health Inspection Service oversees animal and plant health, including preventing the introduction of foreign diseases and pests. USDA's Agricultural Marketing Service establishes quality and marketing grades and standards for a range of agricultural products. USDA's research agencies are also involved in food safety, providing federal funding and collaborating with universities and research institutions.

[2] For more information, see CRS Report RS22600, *The Federal Food Safety System: A Primer*.

[3] FDA's food safety authorities rest primarily in the Federal Food, Drug, and Cosmetic Act (21 U.S.C. §§301 *et seq.*).

[4] Laws governing FSIS include the Federal Meat Inspection Act (21 U.S.C. §§601 *et seq.*), the Poultry Products Inspection Act (21 U.S.C. §§451 *et seq.*), and the Egg Products Inspection Act (21 U.S.C. §§1031 *et seq.*).

[5] FDA's share of the U.S. food supply is approximated by subtracting the reported 10%-20% of foods under USDA's jurisdiction. The 20% estimate is based on information reported by the Government Accountability Office in "Revamping Oversight of Food Safety," prepared for the 2009 Congressional and Presidential Transition. The 10% estimate is based on data from USDA's Economic Research Service on U.S. per capita food consumption at http://www.ers.usda.gov/data/foodconsumption/. See also U.S. Department of Homeland Security, "National Infrastructure Protection Plan: Agriculture and Food Sector Snapshot."

Annual appropriations for FDA and USDA are augmented by existing (currently authorized) user fees. Collected user fees differ between the two agencies: At FDA, user fees authorized under the FDA Food Safety Modernization Act (FSMA, P.L. 111-353) have generated between $10 million and $20 million annually in recent years; at FSIS, authorized user fees have generated between $180 million and $250 million annually (*Figure 2*). Staffing levels also differ between the two agencies: FSIS staff number around 9,200 full-time equivalents (FTEs), while FDA's food-related staff (whose activities extend beyond food safety) number about 3,900 FTEs (*Table 1*).

In addition to setting budgetary amounts, Agriculture appropriations bills may also include policy-related provisions that direct how the executive branch should carry out a specific appropriation. These provisions may have the force of law if they are included in the text of the appropriation, but their effect is generally limited to the fiscal year indicated.

The explanatory statement that accompanies the final appropriation, and the House and the Senate report language that accompanies the committee-reported bills, may also provide policy instructions. These documents do not have the force of law but often explain congressional intent, which the agencies are expected to follow. The committee reports and explanatory statement may need to be read together to capture all of the congressional intent for a given fiscal year. According to the FY2018 Explanatory Statement, "The explanatory statement is silent on provisions that were in both the House Report (H.Rept. 115-232) and Senate Report (S.Rept. 115- 131) that remain unchanged by this agreement, except as noted in this explanatory statement.... The House and Senate report language that is not changed by the explanatory statement is approved and indicates congressional intentions. The explanatory statement, while repeating some report language for emphasis, does not intend to negate the language referred to above unless expressly provided herein."[6]

---

[6] House Committee Print, 115th Congress, Book 1—Consolidated Appropriations Act, 2018.

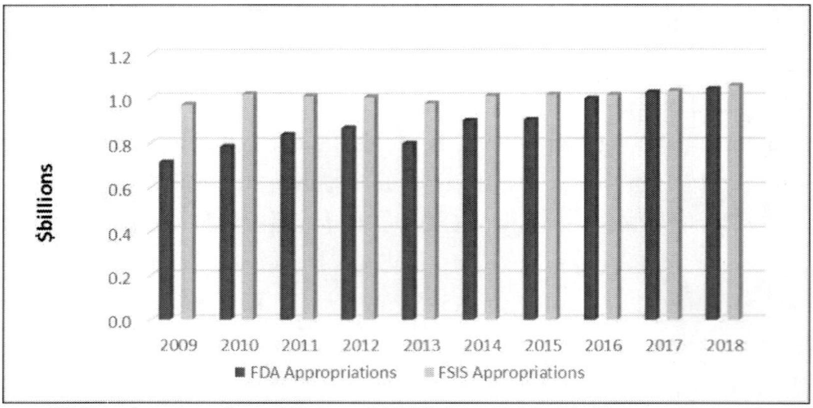

Source: CRS, using data from FDA and FSIS (see Table 1). Data are not adjusted for inflation.

Figure 1. Appropriations for Federal Food Safety Activities, FY2009-FY2018.

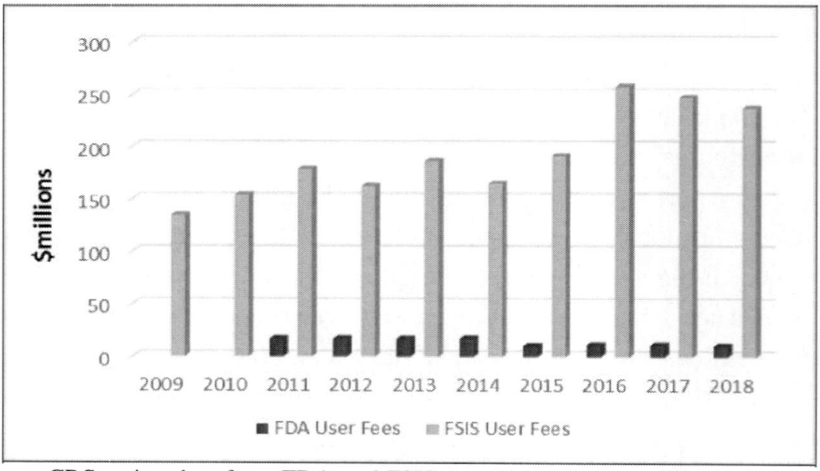

Source: CRS, using data from FDA and FSIS (see Table 1). Data are not adjusted for inflation.

Notes: FDA user fees include food and feed recall fees, food reinspection fees, fees collected under the voluntary qualified importer program, and export certification fees. FSIS user fees include fees for meat, poultry, and egg products inspection; fees for the cost of national laboratory accreditation programs; and trust funds.

Figure 2. User Fees for Federal Food Safety Activities, FY2009-FY2018.

Table 1. Food Safety Enacted Appropriations and Agency-Reported Funding Levels

Agency/Year	Federal Full-Time Equivalents	Enacted Appropriation ($ millions)	Total Agency-Reported Program Level (incl. fees and other funding)[a] ($ millions)
HHS Food and Drug Administration (FDA) "Foods Program"[b]			
FY2009 (P.L. 111-8)	2,995	648.7	NA
FY2010 (P.L. 111-80)	3,387	782.9	NA
FY2011 (P.L. 112-10)	3,605	837.4	NA
FY2012 (P.L. 112-55)	3,611	978.7	NA
FY2013 (P.L. 113-6)c	3,642	796.6	NA
FY2014 (P.L. 113-76)	3,650	900.3	NA
FY2015 (P.L. 113-235)	3,667	903.4	913.8
FY2016 (P.L. 114-131)	3,841	987.3	1,040.8
FY2017 (P.L. 115-31)	3,905	1,025.5	1,025.5
FY2018 (P.L. 115-141)	3,939	1,041.6	1,033.1
FY2019 (President's Budget)	*3,939*	*1,029.9*	*1,040.7*
*FY2019 (*H.R. 5961*)*	*NA*	*1,039.7*	*NA*
FY2019 (S. 2976/*S.Amdt. to* H.R. 6147)	*NA*	*1,052.3*	*NA*
USDA Food Safety and Inspection Service (FSIS)			
FY2009 (P.L. 111-8)	9,343	971.6	1,110.1
FY2010 (P.L. 111-80)	9,401	1,018.5	1,152.6
FY2011 (P.L. 112-10)	9,465	1,008.5	1,168.4
FY2012 (P.L. 112-55)	9,351	1,004.4	1,170.8
FY2013 (P.L. 113-6)[c]	9,158	977.3	1,162.0
FY2014 (P.L. 113-76)	8,933	1,010.7	1,170.0
FY2015 (P.L. 113-235)	8,938	1,016.5	1,206.7
FY2016 (P.L. 114-131)	9,160	1,014.9	1,284.1
FY2017 (P.L. 115-31)	9,243	1,032.1	1,288.0
FY2018 (P.L. 115-141)	9,054	1,056.8	1,266.2
FY2019 (President's Budget)	*9,224*	*1,032.3*	*1,274.1*
*FY2019 (*H.R. 5961*)*	*NA*	*1,049.3*	*NA*
*FY2019 (*S. 2976/*S.Amdt. to* H.R. 6147*)*	*NA*	*1,049.3*	*NA*

Sources: CRS, from enacted appropriations and from annual agency budget justifications for FDA (http://www.fda.gov/AboutFDA/ReportsManualsForms/ Reports/BudgetReports/ default.htm) and FSIS (http://www.obpa.usda.gov/ explan_notes.html). Italicized data and text are not enacted.

Notes: NA = Not available. Data for FDA before FY2015 are not available to make comparisons.

[a] For FDA, reflects available funding for total food safety activities across all FDA programs as well as user fees. For FSIS, includes other federal and nonfederal (fees and trust funds) funding.

[b] Reflects appropriations for Center for Food Safety and Applied Nutrition and related field activities in the Office of Regulatory Affairs. FDA's "Foods Program" includes all food activities, not only food safety.

[c] FY2013 totals represent each agency's FY2013 sequestration operating plans.

FDA Food Safety Activities

FDA's Foods Program covers the agency's food safety activities, as well as certain other food-related programs. The program plays a major food safety role, ensuring that the nation's food supply, quality of foods, food ingredients, and dietary supplements (and also cosmetic products) are safe, sanitary, nutritious, wholesome, and properly labeled. FDA's Foods Program budget accounts for roughly one-third of FDA's total appropriation. FDA's total budget for food safety programs and activities extends beyond the agency's Foods Program to encompass other food and veterinary medicine programs at FDA while also including aspects of other FDA program areas covering food additives, antimicrobial resistance, and nutrition labeling.[7]

Recent Enacted Appropriations

In recent years, congressional appropriators have increased funding for FDA's Foods Program, raising funding from $837.4 million in FY2011 to $1,041.6 million in FY2018. Increased funding over this period followed the enactment of comprehensive food safety legislation in the 111th Congress as part of FSMA. FSMA was the largest expansion of FDA's food safety authorities since the 1930s, and FDA is still actively engaged in implementing the law and its regulations.[8] Over the past few decades, FDA has also had to adapt to the increasing variety and complexity of the

[7] These program areas include portions of the budgets for the Office of Regulatory Affairs, the Center for Food Safety and Applied Nutrition (CFSAN), the Center for Veterinary Medicine, the Office of Veterinary Medicine, and the National Center for Toxicological Research. According to FDA, the agency's budget for food safety activities totals more than $1.2 billion annually. More precise estimates are not readily available or regularly published by FDA.

[8] FSMA amended the Federal Food, Drug, and Cosmetic Act (P.L. 75-717) and did not directly address food safety activities under USDA's jurisdiction. FSMA's provisions mandated new requirements and increased FDA inspections at food facilities, tightened record-keeping requirements, extended oversight to produce farms, mandated product recalls and improvements to foodborne illness surveillance systems, and increased scrutiny of food imports. Information on FDA's implementation of FSMA is provided in CRS Report R43724, *Implementation of the FDA Food Safety Modernization Act (FSMA, P.L. 111-353).*

Since FSMA became law in 2011, congressional appropriators have increased funding for the FDA Foods Program by $204.3 million—an increase of about 24% between FY2011 and FY2018—largely in an effort to support FDA's implementation of FSMA. Currently, FDA's annual appropriation to conduct its food safety oversight activities is roughly similar to that at FSIS (*Figure 1, Table 1*). Cumulative increases to the agency's budget authority to address food safety activities and FSMA implementation since FY2011 have totaled more than $300 million, according to FDA (*Table 2*).

FSMA also authorized an increase in FDA staff, which was expected to reach 5,000 by FY2014.[15] With 3,900 food safety FTEs in FY2018, FDA has fallen short of the goals set forth in FSMA.

Of the total amount appropriated to FDA, the enacted FY2018 act specifically provides that $1.5 million be used to conduct oversight of FDA's programs and operations by HHS's Office of Inspector General and that $1.5 million be used for FDA and USDA to coordinate public education activities regarding crop biotechnology and food and animal feed ingredients derived from biotechnology. The latter allocation originated in the House committee bill.

Congressional appropriations are augmented by existing industry-paid user fees authorized by FSMA. FDA user fees include food and feed recall fees, food reinspection fees, voluntary qualified importer program fees, and fees for certain periodic activities involving reinspection, recall, and export certification.[16] Since FSMA was enacted, these fees have generated between $10 million and $20 million annually (*Figure 2*). The Obama Administration requested authorization of new user fees totaling nearly $170 million annually, but congressional appropriators did not accept its

---

public health groups. See also testimony to the House Appropriations Committee, from Michael Jacobson, Center for Science in the Public Interest, March 20, 2014.

[15] FSMA, P.L. 111-353, §401. By fiscal year, staff level increases were authorized to a goal of not fewer than 4,000 staff (FY2011); 4,200 staff (FY2012); 4,600 staff (FY2013); and 5,000 staff (FY2014).

[16] FSMA, P.L. 111-353, §§107, 401. For other background information, see CRS Report R44309, *FY2016 Appropriations: Selected Federal Food Safety Agencies*.

proposals.[17] User fees are generally established in law by the authorizing committees and not by appropriators.

Table 2. FDA Food Safety Program Funding and Changes to FSMA, FY2011-FY2018

	Total FDA Food Safety Changes	Net FSMA Changes	Explanation
FY2011	+61.3	+59.8	Increases per conference documents provided to FDA
FY2012	+37.3	+39.0	FSMA increases per conference agreement (p. 185)
FY2013[a]	-75.7	-14.6	FSMA increases per S.Rept. 112-163 (p. 77), less permanent base reduction due to sequestration
FY2014	+118.9	+53.5	FSMA increases per S.Rept. 113-46 (p. 79)
FY2015	+26.5	+24.0	Food safety increases per Congressional Record, December 11, 2014(p. H9314)
FY2016	+105.2	+104.4	Explanatory text, Division A, Agriculture, Rural Development, Food and Drug Administration, and Related Agencies Appropriations Act, 2016 (p. 29)
FY2017	+34.9	+35.7	Provided an increase of $38.7 million for food safety activities, of which $35.7 million was to support FSMA implementation
			(Congressional Record, May 3, 2017, Book II, H3334).
FY2018	+12.3	+10.5	According to FDA, increases for various food safety activities specified in report language (e.g., covering FDA oversight activities, lab testing, educational programs, and other activities).
Subtotal	+320.8	+312.3	—

Sources: CRS, from FDA (October-November 2015, with updates in October 2018).

[a] Reflects permanent food safety base reductions due to sequestration. For background information, see CRS Report R43110, *Agriculture and Related Agencies: FY2014 and FY2013 (Post-Sequestration) Appropriations.*

Selected Provisions in the FY2018 Appropriations

Language in the enacted FDA appropriation for FY2018, along with other statements in the House and the Senate committee reports, continues

[17] These proposed fees would have covered the cost associated with food facility registration and inspection, food imports, international couriers, and food contact notifications. See, for example, HHS, "Fiscal Year 2016, Food and Drug Administration, Justification of Estimates for Appropriations Committees," p. 27. For more background information, see CRS Report R44309, *FY2016 Appropriations: Selected Federal Food Safety Agencies*.

to address certain FSMA implementation issues. The enacted FY2018 appropriation act states that "none of the funds made available by this or any other Act" may be used to implement FSMA requirements regarding the regulation of the production, distribution, sale, or receipt of dried spent grain byproducts of the alcoholic beverage production process.[18] Such byproducts are often used as animal feed. As part of FSMA's produce regulations, both the House and the Senate committee reports direct FDA to distinguish between grape varieties intended for wine production and grapes consumed raw. The Senate committee also expressed concerns about FSMA regulations on animal feed derived from cotton ginning and cottonseed.[19] It also directs FDA to provide further clarification to small farms on the requirements for FSMA compliance and to offer technical assistance to facilitate compliance among small farms. The House committee report further states that "it is the intent of Congress for FDA to ensure an even playing field in the application of FSMA regulations as it relates to both domestic and imported producers, processors, and manufacturers of food and animal feed."[20] It also requests that FDA provide the committee with a timeline of all activities associated with the investigation into the illnesses associated with imported pet food.

In addition to FSMA implementation, the enacted FY2018 appropriation addresses other food safety issues under FDA's jurisdiction, including issues regarding fish and seafood safety, the use of biotechnology, standards of identity, and food product labeling. *Table 3* and *Table 4* provide a summary of selected provisions.

On issues regarding fish and seafood safety, for example, both the House and the Senate committees direct FDA to review its final seafood advisory for pregnant and nursing women. The House committee further provides $2.8 million to support the development of appropriate lab methods to detect evidence of seafood decomposition. The Senate committee encourages FDA to increase funding for research into Vibrio illnesses associated with the consumption of raw molluscan shellfish. The

[18] P.L. 115-141, §735.
[19] S.Rept. 115-131, p. 91.
[20] H.Rept. 115-232, p. 65.

enacted FY2018 act prohibits FDA from allowing the "introduction into interstate commerce of any food that contains genetically engineered salmon until the FDA publishes final labeling guidelines."

Both committees direct FDA to continue to monitor fraud in imported olive oil. The Senate committee encourages FDA to respond to a 2012 citizen petition requesting a standard of identity for olive oil and olive pomace oil, while the House committee directs FDA to update its 2014 survey of olive oil products. The House committee report also includes language directing FDA to develop a dairy-specific standard of identity and guidance in accordance with the Dairy Pride Act (H.R. 778/S. 130). It also directs FDA to define *natural* so that there is a uniform national standard for the labeling claims. The Senate committee directs FDA to address concerns about misleading maple syrup marketing.[21]

The enacted appropriation and committee bills also contain other policy riders for FDA's Foods Program that are not necessarily related to the agency's food safety activities. For example, the enacted FY2018 appropriation restricts FDA's ability to deem a food containing partially hydrogenated oils to be unsafe or adulterated and prohibits the agency from using appropriated funds to "develop, issue, promote or advance" regulations or "final guidance applicable to food manufacturers for long-term population-wide sodium reduction."[22] Provisions in the House committee report also address menu labeling requirements and the nutrition facts label, noting that FDA had not issued final guidance regarding the definition of *dietary fiber* and labeling of added sugars.[23]

Selected Provisions in the FY2019 Appropriations

The House-reported bill would appropriate $1,039.7 million for FY2019, which is less than the Senate-passed bill's appropriation of

---

[21] S.Rept. 115-131, p. 95.
[22] P.L. 115-141, §738 and §764, respectively.
[23] H.Rept. 115-232, pp. 72-73.

$1,052.3 million (*Table 1*). By comparison, the Administration's FY2019 budget request recommends $1,029.9 million for FDA's Foods Program.[24] Congressional appropriations would be augmented by FSMA-authorized user fees. The Administration is not proposing any new user fees for FDA's Foods Program.

Of the total amount appropriated to FDA, both the House and the Senate bills would direct agency spending on specific activities. The Senate-passed bill would require that no less than $15 million be used for inspections of foreign seafood manufacturers and field examinations of imported seafood. It would also require that $1.5 million be used to conduct oversight of FDA's programs and operations by HHS's Office of Inspector General. The House-reported bill would also direct $1.5 million for HHS oversight activities and that $3 million be used by FDA and USDA to coordinate public education activities on the safety and benefits of crop biotechnology.

For FDA, the House and Senate FY2019 appropriation bill language, along with other statements in the House and the Senate committee reports, address other food safety issues under FDA jurisdiction (*Table 3*). Similar to those enacted in the FY2018 appropriations act, these include issues addressing FSMA implementation, fish and seafood safety and fraud, the use of biotechnology, standards of identity, and food product labeling. The House and the Senate appropriation and committee bills also contain a number of policy riders for FDA's Foods Program that are not necessarily related to the agency's food safety activities but address certain agency nutritional guidance and regulations. *Table 3* and *Table 4* provide a summary of selected provisions in the proposed FY2019 appropriation bills and committee reports with comparisons with those enacted for FY2018.

---

[24] FDA, "2019 FDA Justification of Estimates for Appropriations Committees, Narrative for FDA Foods," https://www.fda.gov/AboutFDA/ReportsManualsForms/Reports/BudgetReports/ucm617122.htm.

FSIS Food Safety Activities

USDA's FSIS is responsible for inspecting U.S. supplies of meat, poultry, and processed egg products to ensure that they are safe, wholesome, and properly labeled and packaged. The agency's Meat and Poultry Inspection Program conducts continual inspections at federal meat and poultry plants. FSIS also ensures that meat and poultry products imported to the United States are produced under standards equivalent to U.S. inspection standards and facilitates the certification of regulated products.[25]

Selected Provisions in the FY2018 Appropriations

The enacted FY2018 appropriation act for FSIS provided $1,056.8 million (*Figure 1, Table 1*). Compared to FY2011, the FY2018 appropriation for FSIS is nearly $50 million higher.

Of the total amount appropriated to FSIS, the FY2018 act requires that $7.5 million remain available until expended for public health veterinarian recruitment and retention incentives. As in previous years, congressional appropriators direct FSIS to employ no fewer than 148 FTEs during FY2018 that would be dedicated solely to inspections and enforcement related to the Humane Methods of Slaughter Act (7 U.S.C. 1901 *et seq.*).

The FY2018 appropriation for FSIS is divided into various discretionary subaccounts: federal food safety inspection ($943.8 million), state food safety inspection ($61.7 million), international inspections ($16.8 million), and the Public Health Data Communications Infrastructure System ($34.6 million).[26]

---

[25] Examples of FSIS food safety activities include *Salmonella* and *Campylobacter* sampling, foodborne illness outbreak investigations (e.g., *E.coli*, *Salmonella*, and *Listeria*). For more information, see CRS Report RL32922, *Meat and Poultry Inspection: Background and Selected Issues*.

[26] P.L. 115-131, Legislative Text and Explanatory Statement, p. 74. Compared to the FY2017 appropriations, the FY2018 appropriations do not include the Codex Alimentarius

Language in the enacted FSIS appropriation, along with other statements in the House and the Senate committee reports, also addresses a number of concerns. These include certain animal welfare issues, catfish inspection and grading, and restrictions on horse slaughter and imports of chicken from China. *Table 3* and *Table 4* provide a summary of selected provisions.

Regarding catfish inspection, the act requires FSIS to issue equivalence determinations for all countries wishing to continue exporting Siluriformes (catfish) to the United States within 180 days of enactment.[27] The explanatory statement on the enacted FY2018 act (P.L. 115-141) provides that $8 million be made available to fully implement Siluriformes fish and fish product inspection.[28]

Also, as in previous years, the enacted FY2018 appropriations act prohibits FSIS from using funds to inspect horse slaughter facilities or to make use of voluntary inspection fees.[29] Horses are an amenable species under the Federal Meat Inspection Act, and FSIS is responsible for horse slaughter inspection if the horsemeat is for human consumption.

Finally, the enacted FY2018 appropriations act also prohibits the purchase of processed (cooked) poultry meat imported from China for use in the National School Lunch Program, the Child and Adult Food Care Program, the Summer Food Service Program for Children, and the School Breakfast Program.[30] This provision has been added to enacted appropriations bills since FY2015 given continued concerns about China's poor food safety record. The enacted FY2018 act also contains a provision prohibiting the use of funds to finalize USDA's proposed rule regarding China's eligibility to export its poultry products to the United States unless

---

subaccount. In 2017, USDA announced the U.S. Codex Office transferred from FSIS to the Trade and Foreign Agricultural Affairs Mission Area.

27 Catfish inspection was transferred from FDA to USDA in the 2008 and 2014 farm bills (P.L. 110-246, §11016; and P.L. 113-79, §12106). FSIS issued its final rule in December 2015 (80 *Federal Register* 75590, December 2, 2015).

28 P.L. 115-131, Legislative Text and Explanatory Statement, p. 74.

29 P.L. 115-141. §782. Horse slaughter was prohibited in the FY2006 and FY2007 appropriations acts and expanded in the FY2008-FY2011 and FY2014-FY2016 acts to also include a prohibition on voluntary, fee-based inspections. Inspection bans were not in force during FY2012 and FY2013, but no horse slaughter facilities opened during that time.

30 P.L. 115-141. §756.

USDA takes additional steps to ensure the equivalence of China's poultry slaughter inspection system, among other requirements.[31]

Congressional appropriations are augmented by existing user fees (overtime/holiday inspection services) and existing trust fund accounts (voluntary inspection services). In recent years, user fees and other available funds have generated between $180 million and $250 million per year (*Figure 2*), most of which is comprised of existing FSIS user fees.[32]

Selected Provisions in the FY2019 Appropriations

The House-reported bill and the Senate-passed bill would appropriate $1,049.3 million for FY2019 (*Table 1*), roughly $17 million more than the Administration's request.[33]

The Administration is proposing new user fees that would require establishments and official plants to pay fees to cover the costs of federal, state, and international inspection programs for meat, poultry, and eggs. The proposed fee would generate an estimated $660 million.[34] Congressional appropriators have not accepted similar proposals from previous Administrations.

Language in the House and Senate FY2019 FSIS appropriation, along with other statements in the House and the Senate committee reports, address other food safety issues under FSIS jurisdiction (*Table 3*) that are similar to those enacted in the FY2018 appropriations act. These include provisions that address animal welfare, catfish inspection and grading, and restrictions on horse slaughter and imports of chicken from China. The

---

[31] P.L. 115-141, §756. For information on the proposed rule, see 82 *Federal Register* 27625, June 16, 2017. Under equivalency, foreign inspection measures do not have to be the same as in the United States, but they must provide the same level of sanitary and public health protection as U.S. measures. Equivalency for poultry is authorized in the Poultry Products Inspection Act (21 U.S.C. §466).

[32] Based on recent FSIS congressional budget justifications (http://www.obpa.usda.gov/explan_notes.html). Reflects total nonfederal funds, including fees for meat, poultry, and egg products inspection; fees for the cost of national laboratory accreditation programs; and trust funds.

[33] USDA, 2019 President's Budget Food Safety and Inspection Service, p. 22-4.

[34] USDA, 2019 President's Budget Food Safety and Inspection Service, p. 22-15.

House and the Senate committee bills and reports also contain other FSIS provisions. *Table 3* and *Table 4* provide a summary of selected provisions proposed for FY2019 and compare them to those in the FY2018 appropriations act.

A debate on which agency—FSIS or FDA—has regulatory jurisdiction over cell-cultured meat surfaced in early 2018.[35] Both agencies have released public statements claiming oversight of the new technology. The House-reported bill includes a general provision that requires USDA "for fiscal year 2018 and hereafter" to regulate cell-cultured products made from cells of amenable species of livestock, as defined in the Federal Meat Inspection Act (21 U.S.C. §§601 *et seq.*), or poultry, as defined in the Poultry Products Inspection Act (21 U.S.C. §§451 *et seq.*).[36] Elsewhere in the House committee report, however, appropriators acknowledge that the federal jurisdiction of cell-cultured products remains to be determined.[37] In November 2018, a joint statement from FDA and FDA announced that both agencies "should jointly oversee the production of cell-cultured food products derived from livestock and poultry."

PROVISIONS IN BILL TEXT AND REPORT LANGUAGE

Table 3 compares selected policy provisions that have been identified in the Agriculture Programs (Title I), the Related Agencies and Food and Drug Administration (Title VI), and General Provisions (Title VII) titles of the FY2018 Agriculture appropriations act and the FY2019 Agriculture appropriations bills related to federal food safety activities at both FDA and FSIS. Many of these provisions have been included in past years' appropriations acts.

---

[35] Cell-cultured meat (also referred to as lab-grown meat, clean meat, in vitro meat, imitation meat, synthetic meat, and fake meat) is grown in laboratories from animal cell cultures. For more information, see CRS In Focus IF10947, *Regulation of Cell-Cultured Meat.*

[36] H.R. 5961, §736.

[37] H.Rept. 115-706, p. 183.

Table 3. Comparison of Selected Appropriations Provisions in the Enacted FY2018 Appropriations Act and House and Senate Bill Text for FY2019

FY2018	FY2019	
Enacted (P.L. 115-141)	House-Reported (H.R. 5961)	Senate-Passed (H.R. 6147)
FDA Provisions		
FSMA Regulations		
Restricts FDA funds from being used to implement FSMA regulation of the production, distribution, sale, or receipt of	Similar to P.L. 115-141 (§733).	Similar to P.L. 115-141 (§735).
Enacted (P.L. 115-141)	House-Reported (H.R. 5961)	Senate-Passed (H.R. 6147)
dried spent grains from the alcoholic beverage production process (§735). (See S.Rept. 115-131; H.Rept. 115-232.)		
No comparable provision.	Restricts FDA funds from being used to enforce its final produce standards with respect to regulating the production, distribution, sale, or receipt of grape varietals that are grown, harvested, and used solely for wine (§755).	Similar to House provision (§753).
GE Salmon		
Prohibits the introduction into interstate commerce of foods that contains genetically engineered salmon until labeling guidelines are finalized (§770).	Establishes disclosure requirements related to GE salmon or other finfish in accordance with USDA rules (§766).	Similar to P.L. 115-141 (§740).
Fish Advisories		
No comparable provision.	No comparable provision.	Directs FDA to provide certain advice regarding its notice (along with the U.S. Environmental Protection agency) about eating fish (82 *Federal Register* 6571) (§764).

Table 3. (Continued)

FY2018	FY2019	
Directed Funding Allocations		
Of available funds, requires $1.5 million for HHS's Office of Inspector General for oversight of FDA programs and operations (P.L. 115-141, FDA general text).	Similar to P.L. 115-141 (FDA general text).	Similar to P.L. 115-141 (FDA general text). See also S.Rept. 115-259.
Of available funds, requires $1.5 million for FDA and USDA coordinated public education activities on the safety and benefits of crop biotechnology and food and animal feed ingredients derived from biotechnology (P.L. 115-141, Explanatory Notes, p. 85).	Similar to P.L. 115-141 (FDA general text), but increases amount to $3 million.	No comparable provision.
No comparable provision.	No comparable provision.	Of available funds, requires $15 million be used for inspections of imported seafood (FDA general text).
Selected Policy Riders		
Directs FDA not to deem partially hydrogenated oils as unsafe or unadulterated (§738).	Similar to P.L. 115-141 (§740).	Similar to P.L. 115-141 (§736).
Restricts FDA funds from being used to develop, issue, promote, or advance sodium regulations (§764).	Similar to P.L. 115-141 (§752).	Similar to P.L. 115-141 (§750).
Other Selected General Provisions		
Directs FDA, and other federal agencies, to submit to appropriators a detailed spending plan by program, project, and activity for all funds made available (P.L. 115-141, §723).	Similar to P.L. 115-141 (§722).	Similar to P.L. 115-141 (§723).
FSIS Provisions		
Federal Meat, Poultry, and Egg Inspection		
Directs FSIS to employ no fewer than 148 FTEs during FY2018 dedicated solely to inspections and enforcement related to the Humane Methods of Slaughter Act (P.L. 115-141, FSIS general text).	Similar to P.L. 115-141 (FDA general text).	Similar to P.L. 115-141 (FSIS general text).

FY2018	FY2019	
Prohibits imports from China of raw or processed poultry products for a number of school lunch and food programs (§756). Restricts the use of funds to finalize USDA's proposed rule regarding China's eligibility to export its poultry products to the United States (§760).	Includes similar language to P.L. 115-141 (§748 and §751).	No comparable language.
Prohibits funds from being used for horse slaughter and inspection (§782).	No comparable language.	Similar to P.L. 115-141 (§758).
Catfish Inspection		
Requires that FSIS issue equivalence determinations for all countries wishing to continue exporting Siluriformes to the United States (P.L. 115-141, p. 74).	No comparable language.	Directs FSIS to continue implementing regulations on catfish inspection and grading (FSIS general text).
Country Audits		
Directs USDA to conduct country audits related to, among other things, animal disease control and surveillance, lab capabilities, preparedness, and response (§742).	No comparable language.	No comparable language.
Enacted (P.L. 115-141)	House-Reported (H.R. 5961)	Senate-Passed (H.R. 6147)
Products from Cells of Amenable Species		
No comparable language.	Directs FSIS to regulate products made from cells of amenable species of livestock (§736).	No comparable language.
Veterinarian Recruitment and Retention		
Of available funds, requires $7.5 million remain available until expended for public health veterinarian recruitment and retention (P.L. 115-141, p. 74).	No comparable language.	No comparable language.

Source: CRS.

Table 4 compares selected policy provisions in the FY2018 and the FY2019 Agriculture appropriations report language or explanatory statements related to federal food safety activities at both FDA and FSIS. Many of these provisions have also been included in past years' appropriations acts.

Table 4. Selected Appropriations Provisions in Report Language, FY2018 and FY2019

FY2018	FY2019	
Accompanying Report Language (S.Rept. 115-131, H.Rept. 115-232, and Explanatory Statement)	House Committee Report (H.Rept. 115-706)	Senate Committee Report (S.Rept. 115-259)
FDA Provisions		
FSMA's Regulatory Requirements/Assessment		
Directs FDA to distinguish between grape varieties for wine production and grapes consumed raw (S, H).	No comparable provision, but see H.R. 5961, §755.	No comparable provision, but see H.R. 6174, §754.
Directs FDA to continue its progress in improving federal oversight and monitoring of state inspection programs, reviewing and strengthening internal directives and processes, and identifying new methods to improve oversight capabilities, including modernizing IT systems and infrastructure (H).	No comparable provision.	No comparable provision.
Encourages FDA to fully use the Centers for Food Safety and Nutrition Centers of Excellence to support research of FSMA activities (S).	No comparable provision.	Similar to language in S.Rept. 115-131.
Expresses concerns about regulations on byproducts of cotton ginning and cottonseed for animal feed use (S).	No comparable provision.	Similar to language in S.Rept. 115-131.
Provides sufficient monies to fund the FDA's Food Contact Notification Program but does not include proposed user fees (H). Recommendations do not include proposed user fees (S).	Provides sufficient funds for FDA's Food Contact Notification Program.	Recommends that proposed Food Contact Notification user fees not be included.
Directs FDA to further clarify compliance by small farms and urges FDA to communicate via guidance and technical assistance (S).	No comparable provision.	Similar to language in S.Rept. 115-131.
Imported Foods		
Requests FDA provide a timeline of all activities associated with the investigation into the illnesses associated with pet food imports. Also requests FDA provide semi-annual reports on the status of the investigation until the issue has been resolved (H).	Requests that FDA provide a timeline of all activities associated with its investigation into pet illnesses associated imported pet food.	Recommends increases of $2.8 million for food import safety activities (as part of general increases for food safety activities).
Maintains FY2017 funding of $7.5 million	Directs FDA to spend $7.5	Directs FDA to provide an

FY2018	FY2019	
Accompanying Report Language (S.Rept. 115-131, H.Rept. 115-232, and Explanatory Statement)	House Committee Report (H.Rept. 115-706)	Senate Committee Report (S.Rept. 115-259)
for FDA's Office of Global Regulatory Operations and Policy to enhance the compliance and verification of foreign high- risk manufacturers and exporters of food and other products (H, S).	million of FY2018 appropriations for foreign high-risk inspections.	update on efforts to address foreign high- risk inspections.
Directs FDA to submit a report on the potential for implementing pilot programs allowing for public-private partnerships at high-volume ports of entry (H).	No comparable provision.	No comparable provision.
Directs FDA to work with local governments at high- volume ports of entry to reduce the risk of food-borne illnesses and improve local capacity (H). States it is the sense of the committee that FDA, USDA, and Customs and Border Protection should reassess staffing models to ensure available workforce at U.S. ports of entry (H).	Directs FDA to work with local governments at high-volume ports of entry to reduce the risk of food-borne illnesses and report back to the committee.	No comparable provision.
Working with States and Other Partners		
Recommends FDA consider suggestions from its public hearing ("Strategic Partnerships to Enhance the Safety of Imported Foods') regarding the use of private-sector third-party auditors to help	Encourages FDA to enter into MOUs with government food safety programs. Provides $5 million for	Recommends increases of $7.2 million for FSMA cooperative agreements and $5 million to address food safety outbreaks (as part of
implement FSMA (S). Encourages FDA to work in partnership with existing government food safety programs through memorandums of understanding (MOUs) to verify compliance with FSMA to rules once they are finalized as a way to eliminate duplication and provides $5 million for the Food Safety Outreach Program under the National Institute of Food and Agriculture (serving as the sole agency providing food safety training, education, outreach, and technical assistance at the farm level) (H). Directs FDA to work with any state that designates a new state implementing agency	USDA's Food Safety Outreach Program. Directs FDA to continue its FSMA education and outreach to domestic and foreign producers, processors, and manufacturers. Urges FDA to provide sufficient resources to state education/ inspection programs.	general increases for food safety activities). Directs FDA to work with any state that designates a new implementing agency to ensure it receives funding under existing cooperative agreements.

T

U

Table 4. (Continued)

FY2018	FY2019	
to ensure it continues to receive funding under existing cooperative agreements (S).		
Seafood Safety		
Provides $2.8 million to help develop lab methods to detect evidence of seafood decomposition (E, H).	No comparable provision.	No comparable provision.
No comparable provision.	No comparable provision.	Urges FDA to complete single lab validation and testing for detecting brevetoxins associated with neurotoxic shellfish poisoning in molluscan shellfish.
No comparable provision.	No comparable provision.	Of monies available for the National Antimicrobial Resistance Monitoring System, directs FDA to use $0.5 million to test antibiotic resistance in imported seafood.
Encourages FDA to increase funding for research into Vibrio illnesses associated with the consumption of raw molluscan shellfish (S).	No comparable provision.	Similar to language in S.Rept. 115-131.
Directs FDA to review its seafood advisory for pregnant and nursing women (S, H).	Directs FDA to post nutrient values of fish consumption during pregnancy on its website.	Directs FDA to reissue its final "Advice About Eating Fish" (82 *Federal Register* 6571).
Standards of Identity; Food Fraud/Defense		
Directs FDA to develop a dairy-specific standard of identity and guidance in accordance with the Dairy Pride Act, H.R. 778/S. 130 (H).	Directs FDA to develop a standard of identity for dairy products and issue guidance on how to implement and enforce the standard.	No comparable provision.
Directs FDA to monitor olive oil import fraud (S, H), update its 2014 olive oil survey (H), and respond to a 2012 citizen petition requesting a standard of identity for olive oil and olive pomace oil (S).	Directs FDA to establish a separate U.S. standard of identity for different grades of olive oil.	Directs FDA to establish a standard of identity for different grades of olive oil and olive-pomace oils.
Urges FDA to engage in further collaborative dialogue with stakeholders regarding its final rule "Mitigation	No comparable provision.	Encourages FDA to work with businesses to identify food defense practices that

FY2018	**FY2019**	
Strategies to Protect Food Against Intentional Adulteration'' (81 Federal Register 34165) (E).		effectively protect public health.
Directs FDA to define natural and establish a uniform national labeling standard (H).	No comparable provision.	No comparable provision.
Of the amount provided for CFSAN, requires an increase of $3 million for the Office of Nutrition and Food Labeling to prioritize efforts regarding standards of identity and related product labeling (H).	Similar to language in H.Rept. 115-232.	No comparable provision.
Directs FDA to address concerns about misleading maple syrup and product marketing (S).	No comparable provision.	Similar to language in S.Rept. 115-131.
Other Selected Provisions		
Encourages the National Academies of Sciences, Engineering, and Medicine to conduct a study on new technologies to promote microbiological food safety and prevent foodborne illnesses (H).	No comparable provision.	No comparable provision.
Directs FDA to consider all risk assessments before finalizing the guidance and policy on Listeria Monocytogenes to ensure it is risk-based and reflects good and achievable industry practice (H).	Similar to language in H.Rept. 115-232.	Urges FDA to complete a comprehensive risk assessment before changing its guidance regarding Listeria Monocytogenes in ready-to-eat foods.
No comparable provision.	Provides $3 million for USDA and FDA to educate the public on the safety and benefits of crop biotechnology and food and animal feed ingredients derived from biotechnology.	No comparable provision.
Encourages FDA and USDA to provide outreach and guidance to food manufacturers and retailers on food date labeling (H).	Similar to language in H.Rept. 115-232.	No comparable provision.
Directs FDA to comply with Title 31 of the U.S. Code regarding the development of its organizational priority goals and outcomes (e.g., performance outcome and output measures, and also efficiency and customer service measures) (H).	Similar to language in H.Rept. 115-232. Also directs FDA to provide public information on the link between FSMA activities and performance measures, especially public health outcome measures.	No comparable provision.

Table 4. (Continued)

FY2018	FY2019	
Accompanying Report Language (S.Rept. 115-131, H.Rept. 115-232, and Explanatory Statement)	House Committee Report (H.Rept. 115-706)	Senate Committee Report (S.Rept. 115-259)
No comparable provision.	No comparable provision.	Urges FDA to dedicate extra personnel to speed the review and approval process regarding animal feed ingredients.
Maple Syrup Labeling		
Directs FDA to report to the committee on its efforts to implement regulations and provide clarity to the maple syrup and honey industries on the labeling of the sugar content of their packaged products (S).	No comparable provision, but see H.R. 5961, §764.	Directs FDA to work with maple syrup and honey sectors on labeling for single-ingredient products where sugar is naturally occurring (not added). See also H.R. 6174, §768.
Selected Policy Riders, Other Guidance		
Directs FDA to extend the compliance date for manufacturers for the Nutrition Facts Label Final Rule and the Serving Size Final Rule and finalize these regulations before July 26, 2018 (E). Directs FDA to issue guidance on pending dietary fiber ingredients (E). Addresses menu labeling requirements and nutrition facts labeling, including guidance regarding the definition of dietary fiber and labeling of added sugars (S, H).	Urges FDA to reduce the burden on and add flexibility for businesses under the menu labeling rule. Directs FDA to allow three years for industry compliance after final guidance is issued regarding dietary fiber sources.	Includes language regarding dietary fiber similar to language in S.Rept. 115-131. Directs FDA to report on activities and resources spent on nutritionrelated activities of CFSAN and associated field offices.
FSIS Provisions		
Humane Methods of Slaughter		
Directs FSIS to provide an annual report on the implementation of objective scoring methods undertaken to enforce the Human Methods of Slaughter Act (S, H).	Directs FSIS to ensure that inspectors comply with the Humane Methods of Slaughter Act.	Directs FSIS to ensure that inspectors comply with the Humane Methods of Slaughter Act. Requires annual reporting.
Siluriformes (or Catfish) Inspection		
Provides $8 million to fully implement Siluriformes fish and fish product inspection (E). Directs FSIS to complete equivalence determinations for all countries wishing to continue exporting Siluriformes to the United States. In addition, directs	Restates FSIS is expected to comply with the FY2018 enacted appropriation in determining equivalency for countries wishing to export catfish to the United States.	No comparable language.

FY2018	**FY2019**	
Accompanying Report Language (S.Rept. 115-131, H.Rept. 115-232, and Explanatory Statement)	House Committee Report (H.Rept. 115-706)	Senate Committee Report (S.Rept. 115-259)
FSIS to complete equivalence determinations on a country-by-country basis based on volume of catfish exports to the United States (S, H).		
Cultured Meat		
No comparable language.	Acknowledges regulatory responsibility is still being determined for cell-cultured meat. (See "Minority views of the Hon. Nita Lowey and the Hon. Sanford D. Bishop, Jr."). See also H.R. 5961, §736.	No comparable language.
Breed Claim Standards		
Direct FSIS to (1) evaluate current processes for the labeling of 100% pure breed claim standards for livestock products and (2) identify instances of conflicting 100% pure breed claim		No comparable language.
standards for livestock (S).		
Game Birds		
Directs FSIS to provide a report on the impact of defining games birds (e.g., quail) as amendable in federal code regarding inspection (S).	No comparable language.	No comparable language.

Source: CRS.

Notes: Provisions in the enacted FY2018 Explanatory Statement are cited as (E) and available at U.S. Congress, House Committee on Appropriations, Committee Print on H.R. 1625/P.L. 151-141, Book 1—Consolidated Appropriations Act, 2018, 115th Cong., 2nd sess., 29-456. Provisions listed in the House report (H.Rept. 115-232) are cited as (H). Provisions listed in the Senate report (S.Rept. 115-131) are cited as (S).

INDEX

A

D

E

F

G

H

I

J

L

M

N

O

P

R

S

V

W

Agricultural Developments and Policy Implications

Editors: Abraham Porter

Series: Agriculture Issues and Policies

Book Description: This book is a compilation of CRS reports on agricultural developments and policies. Some topics discussed herein include industrial hemp, federal agricultural disaster assistance programs and the 2014 farm bill.

Softcover ISBN: 978-1-53614-491-8
Retail Price: $95

Agribusiness: Management, Opportunities and Challenges

Editor: Ronald Luna

Series: Agriculture Issues and Policies

Book Description: The issue of how to assess the sustainability of governance structures is one of the most topical for researchers, farmers, investors, administrators, policymakers, interest groups, and the public at large.

Softcover ISBN: 978-1-53614-413-0
Retail Price: $95